University of Massachusetts Lowell Libraries

AF697233

Sill/Zörner

Steam Turbine Generators Process Control and Diagnostics

Steam Turbine Generators Process Control and Diagnostics

Modern Instrumentation for the Greatest Economy of Power Plants

by Udo Sill and Walter Zörner

Publicis MCD Verlag

Die Deutsche Bibliothek – CIP-Einheitsaufnahme

Sill, Udo:
Steam turbine generators process control and diagnostics : modern instrumentation for the greatest economy of power plants / von Udo Sill und Walter Zörner. [Siemens]. – Erlangen ; München : Publicis-MCD-Verl., 1996

ISBN 3-89578-040-5

NE: Zörner, Walter:

process control and

TJ 735 .S52 1996
Sill, Udo.
Steam turbine generators
process control and

ISBN 3-89578-040-5

Editor: Siemens Aktiengesellschaft, Berlin and Munich
Publisher: Publicis MCD Verlag, Erlangen

© 1996 by Publicis MCD Werbeagentur GmbH, Verlag, Munich
This publication and all parts thereof are protected by copyright. All rights reserved. Any use of it outside the strict provisions of the copyright law without the consent of the publisher is forbidden and will incur penalties. This applies particularly to reproduction, translation, microfilming or other processing, and to storage or processing in electronic systems. It also applies to the use of individual illustrations and extracts from the text.

Printed in Germany

Preface

In order to achieve the highest possible levels of power plant availability and economy, it is essential that facilities are operated in a prudent, thermodynamically optimum fashion with cost-effective maintenance. In reaching this goal, a high degree of instrumentation and control systems are vital for today's state-of-the-art plant. The high degree of instrumentation and control systems at today's power plants produce extensive volumes of data, and thus place extreme demands on operating personnel with regards to process monitoring and control, particularly in terms of transient procedures and potential malfunctions.

Diagnostics systems augment the monitoring of relevant process variables (recognition and evaluation of deviations from design conditions) and facilitate gaining a comprehensive overview of process relationships. Moreover, considerable savings can be achieved by applying diagnostics systems for thermodynamic optimization, to prevent secondary damage, and to optimizing maintenance strategies. To ensure achieving cost-effective, optimized plant operation, Siemens/KWU has developed state-of-the-art I&C and diagnostics systems which can efficiently monitor, control, optimize processes and recognizing potential malfunctions, while simultaneously providing technical support to operating personnel.

Backed by extensive process engineering experience and in-depth power plant operational know-how in the fields of plant engineering, process automation and I&C systems, Siemens/KWU instrumentation, control and diagnostics systems are ideally suited for a wide range of power plant applications.

This book was written for power plant planning engineers, operators, consultants, and others active in this field. The articles in this book have been compiled to portray a broad picture of Siemens/KWU's state-of-the-art steam turbine and turbinegenerator series, I&C equipment and diagnostics systems. As it is not meant to be a textbook, discussions here have been limited to include only those process engineering fundamentals and software concepts needed to understand the function of these systems. Primary emphasis is thus placed on engineering details of common interest. Due to recent technological developments, a good deal of space has been devoted to diagnostics systems.

Erlangen, November 1996

Siemens Aktiengesellschaft
Power Generation Group (KWU)

Contents

1 **Introduction** 11

2 **Overview of the Various Power Plant Design Concepts** 13

2.1 Steam Turbine and Turbogenerator Series 14
2.1.1 Steam Turbines 14
2.1.2 Turbogenerators 17

2.2 Steam Turbine Generator Sets 17
2.2.1 Rating Turbine Generator Series 17

2.3 Operating Mode of Steam Turbines 21
2.3.1 Definition of the Load Condition for Steam Turbines with Control Stage 24
2.3.2 Definition of the Load Condition for Steam Turbines with Full-Arc Admission 25

2.4 Auxiliary Systems of the Steam Turbine Generator Set. 26

3 **Overview of the Process Control and Information System TELEPERM XP** 27

3.1 Automation Systems for Process Automation 27
3.1.1 Subsystems 28

3.2 Bus System SINEC H1 29

3.3 Process Control and Information System OM 650. 29

3.4 Engineering System ES 680 29

3.5 I&C Concepts 30

3.6 System Test. 32

4 **Steam Turbine Generator Set Monitoring, Control and Protection Systems** 33

4.1 Overview of the Monitoring System 33

4.2 Overview of the Control and Protection Systems 34

4.3 Steam Turbine Controller 37
4.3.1 Controller Structure 38

4.4 Stop and Control Valves for Main and Reheat Steam 42
4.4.1 Electrohydraulic Actuator for Steam Valves (EHA) 43
4.4.2 Electrohydraulic Actuator Control Unit 45
4.4.3 Automatic Turbine Valve Tester 46

4.5 Protection System of the Steam Turbine Generator Set 47
4.5.1 Turbine Protection Criteria 47
4.5.2 Generator Protection Criteria 50
4.5.3 Fire Protection for the Steam Turbine Generator Set 53
4.5.4 Structure of the Turbine Generator Protection Equipment 53
4.5.5 Turbine Trip System 54

4.6 Service Life Control 54
4.6.1 Start-Up of the Steam Turbine Set (Low-Alloyed Material) 55
4.6.2 Start-Up and Load Changes for High-Alloyed Materials 58
4.6.3 Turbine Stress Controller (TSC) 59

4.7 Seal Steam and Gland Sealing System 66
4.7.1 Seal Steam System 66
4.7.2 Gland Sealing System 67

4.8 Bypass Station 67
4.8.1 HP Bypass Station 68
4.8.2 LP Bypass Station 69
4.8.3 LP Bypass Controller 70

4.9 Turbine Drain System 71

4.10 Oil Supply System 72
4.10.1 System Configuration 72
4.10.2 Oil Supply during Turbine Generator Set Operation 74

4.11 Generator Coolant Temperature Controller for Generators with Water-Cooled Stator Windings 75

4.12 Excitation System 76
4.12.1 Brushless Excitation 76
4.12.2 Static Excitation 76
4.12.3 Generator Voltage Regulator 76

4.13 Open-Loop Control at the Turbine Generator Unit 79
4.13.1 Functional Groups 80
4.13.2 Function of the Open-Loop Control 81

5 Diagnostics System DIGEST 83
5.1 Introduction 83
5.2 Basic Concept 83
5.3 Structure of DIGEST 87

5.4	Modular Architecture	87
5.4.1	System Framework	89
5.5	Software Application	90
5.5.1	Expert Systems (XPS)	90
5.5.2	UNIX Operating System	95
5.5.3	Neural Networks	96
5.5.4	Genetic Algorithm	97
5.6	Software Features and Graphics	99
5.6.1	Software Features	99
5.6.2	Graphics	101
6	**Integration of DIGEST into the Instrumentation and Control System**	104
6.1	Conceptual Features	104
6.2	Software Aspects and Hardware	105
7	**DIGEST Modules in Application**	107
7.1	Vibration Analysis and Diagnosis Module	107
7.1.1	Analysis Concept	109
7.1.2	Diagnostics Concept	120
7.2	Service Life Evaluation Module	122
7.3	Thermodynamics Modules	122
7.3.1	Thermodynamics Modules for Fossil-Fired Power Plants	122
7.3.2	Thermodynamics Module for Combined-Cycle Power Plants	127
7.3.3	Unit Commitment Module	130
7.4	Windage Module (Avoidance of Increased Exhaust Steam Temperature in HP and LP Turbine Sections)	137
7.4.1	Thermodynamic Variables and Operating Data Influencing Blading Windage	137
7.4.2	Concept of the Windage Module	140
7.5	Blade Vibration Information Module	144
7.5.1	Functional Description	145
7.6	Condenser Analysis Module	151
7.6.1	Thermodynamics	151
7.6.2	Air Extraction Units	154
7.6.3	The Condenser Analysis Sequence	155
7.7	Bar Temperature Analysis Module for Generators with Water-Cooled Stator Windings	157
7.7.1	Structure of the System	158
7.7.2	Analysis Method	160

7.8 Monitoring (Analysis) and Diagnostics Module for Early Fault Detection on Electrical Equipment 162
7.8.1 Monitoring (RF Analysis) Module 162
7.8.2 Diagnostics Module (RF Diagnosis) 166
7.9 Diagnostics Module for Detecting Hydrogen Leaks on Generators (H2-Sealing) 172
7.9.1 Structure of the System 172
7.9.2 Analysis and Diagnosis 174
7.10 Diagnostics Module for Recognizing High Humidity-Conditions and Water Losses in Generators 178
7.10.1 Structure of the System (Example THFF Series) 179
7.10.2 Analyses and Diagnoses 181
7.11 Diagnostics Module for Recognizing Disturbed Cooling of Generators with Water-Cooled Rotor and Stator 182
7.11.1 Structure of the System (THFF Series) 183
7.11.2 Analyses and Diagnoses 186
7.12 Gas Turbine Thermodynamics and Fault Analysis Module 186
7.12.1 Analyses and Thermodynamic Calculations 187
7.12.2 Data Processing and Hardware 191
7.13 Outlook of Anticipated Modules 192
7.13.1 Bearings 192
7.13.2 Turbine Casings 192
7.13.3 Valves 193
7.13.4 Excitation 193
7.13.5 Feed Water Chemistry 193

8 Benefit of DIGEST 194

8.1 Thermodynamically Optimum Mode of Operation 197
8.2 State-of-the-Art Methods for Fault Detection 198
8.3 Monitoring of Service Life Expenditure 205
8.4 Maintenance 206

9 Service and Experience with DIGEST 212

9.1 Service, Basic Concept 212
9.2 Service, System Upkeep 212
9.3 Experience with DIGEST Modules 213

Literature 215

1 Introduction

The discovery of the electrodynamic principle by Werner von Siemens in 1886 laid the foundations for the generation of electricity which ultimately led to the reliable and extensive supply of electrical power to the consumer for a wide range of purposes. Economic utilization of the energy generated using steam in turbomachinery became a reality about 100 years ago when de Laval and Parsons built the first steam turbines.

As industrialization progressed at the turn of the century, the steam turbine replaced the piston engine. The realization that steam turbine generators could be used to convert large amounts of energy in a confined space ensured that they were soon in use throughout the world.

In the knowledge that steam turbine generators are a key component in the generation of electrical energy, Siemens & Schuckert introduced steam turbine construction in 1927. Since then, Siemens continued to advance and influence the development of steam turbine generators. Some important milestones in the development of power plant construction at Siemens were the installation of the largest 3000 rpm steam turbine generator in the world at the Schelle power plant in 1934, a position achieved as market leader of steam turbine manufacturers by the application of highest main steam pressures and temperatures for steam turbines in the 1950's, installation of the largest nuclear power plant in the world, the Biblis nuclear power plant operated by the utility Rheinisch Westfälische Elektrizitätswerke AG (RWE), in 1976, the largest 3,000 rpm conventional steam turbine generator with 828 MW in Germany for Preussen Elektra in the Heyden power plant in 1987 and, most recently the commission to build a 1,000 MW steam turbine generator with the highest steam parameters for RWE.

In 1969, the Power Plant Engineering Divisions of AEG and Siemens joined forces and this led to the founding of KWU and later to the full integration of this division into Siemens. The pooling of the know-how and experience of these two companies resulted in a number of innovative developments in power plant engineering.

The continuous rise in fuel costs for fossil-fired power plants and pollution control measures led to the construction of power plants with complex thermal cycles, such as fully-fired and parallel-powered combined-cycle power plants to achieve the greatest possible levels of efficiency. Many fossil-fired power plants also utilize heat extraction for district heating and other processes. The challenges of power plant engineering are met by building power plants that are optimally designed for these cycles: the engineering services provided by Siemens are geared toward meeting special demands of fossil-fired power plants of all types. Measures aimed at optimizing energy conversion and increasing energy utilization have always been the focal point of our efforts and have led to a considerable increase in the degree of efficiency, therefore contributing to the conservation of resources and the conservation of the environment.

Siemens/KWU steam turbines meet the specific requirements of all types of fossil-fueled and nuclear power plants. Measures geared toward increasing efficiency have been our highest priority for many years and have met with considerable success. Our steam turbines are known for their sturdy construction, long operating life, and serviceability. The steam turbines are manufactured in a number of different countries in Siemens own production facilities or under license by various partners.

The introduction of instrumentation and control (I&C) in power plants has also had a marked influence on improving the cost-effectiveness of power plants. For this reason, we have continuously developed our I&C concepts with the aim of optimizing power plant processes to the fullest, and increasing the reliability and service life of power plants. The experience we have accumulated in equipping over 1000 power plants of many different types and sizes with I&C equipment and the know-how that we have in this field make us one of the leading and most competent power plant suppliers. Modern process I&C which is based on high-performance microprocessors has made it possible for us to introduce modern digital process control systems from a modular building block system that also fulfills all the demands for a high degree of reliability, and simple handling and maintenance.

With regard to the cost-effectiveness and the service life of power plants, our technical approach is therefore oriented toward these quality demands. The certificates DIN ISO 9001 awarded to each of our divisions confirm that their work processes meet standardized QM system requirements and confirm their proficiency in these work processes.

The aim of modern power plant I&C is the greatest degree of plant safety and availability while at the same time ensuring optimum cost-effectiveness. For such tasks we now have available diagnostics systems, that are helpful and promising tools with a record of successful field applications. It is imperative that the diagnostics system be an integral part of the I&C system. This can only be achieved with a holistic system concept that embraces every component from the sensor to the man-machine interface. Early detection and localization of disturbances mean fast response so that maintenance and repair work can be planned in good time. Unexpected and expensive downtimes no longer occur.

During the preparation of this book colleagues in their respective specialized fields gave us their generously support and assistance in checking particular sections. At this opportunity we would like to express our gratitude to these colleagues.

2 Overview of the Various Power Plant Design Concepts

In the past, the continually increasing cost of fossil fuels and the capital investment in environmental protection technology for fossil-fired power plants in the Federal Republic of Germany have been a driving force behind efforts to improve thermal power plant efficiency. Extensive research and development in the field of steam turbine design has led to a substantial improvement in thermal performance. The maximum net efficiency which is theoretically possible has almost been reached; a further increase in generating efficiency is only possible through optimization of details. Special emphasis is being placed on improving the operating availability of these turbines by technically feasible means with the aim of enhancing overall power plant economics [1].

Plant suppliers and operators have cooperated in developing various power plant design concepts aimed at improving power plant economics and reducing pollutant emissions, concepts which in the future will compete with the proven steam power plant design using atmospheric pulverized-coal firing or other energy sources such as natural gas and oil. High efficiencies can be achieved by using currently available concepts for fossil-fired power plants as well as some concepts rapidly approaching technical and commercial feasibility. Fossil-fired concepts available today include the following:

- ▷ Fully-fired and parallel-powered combined-cycle power plants operating on natural gas and/or coal and incorporating gas turbine topping cycles upstream of the steam generators,
- ▷ Unfired combined-cycle power plants firing natural gas and/or oil,
- ▷ Unfired combined-cycle power plants with integrated coal gasification,
- ▷ Combined-cycle power plants employing pressurized fluidized bed boilers,
- ▷ Coal-fired steam power plants all types (with heat extraction for cogeneration).

The above concepts will be complemented in the future by power plant technologies currently under development such as combined cycles based on pressurized pulverized-coal combustion. For nuclear power stations proven design concepts are available. In order to fulfill the requirements for the different types of power stations, Siemens/KWU is in the position to offer steam turbine generator sets for each of these applications.

Series	Basic Configuration	Speed rpm	Capability Range MW	Max. Operating Steam Conditions bar	°C
E Series without reheat	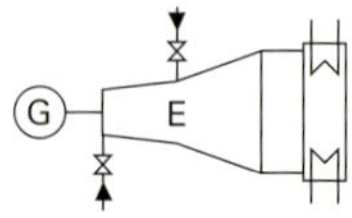	3,000	50 - 300	80	540
		3,600	30 - 200		
DN Series without reheat	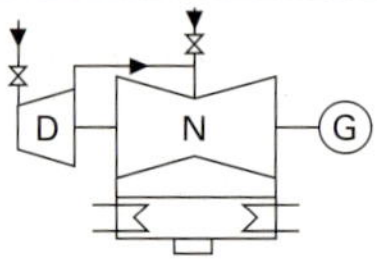	3,000	50 - 300	80	540
		3,600	30 - 200		
HE Series with reheat	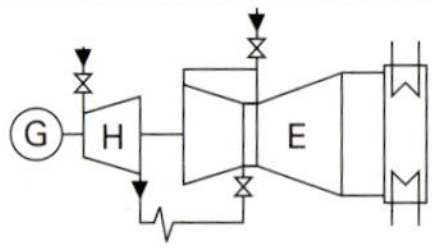	3,000	50 - 400	300	600/600
		3,600	30 - 400		
KN Series with reheat	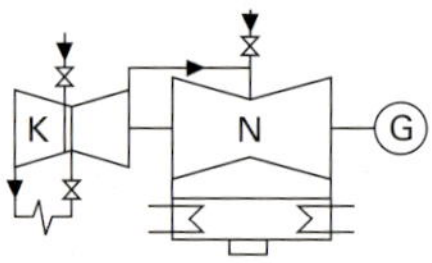	3,000	50 - 450	170	540/540
		3,600	30 - 450		
HMN Series with reheat	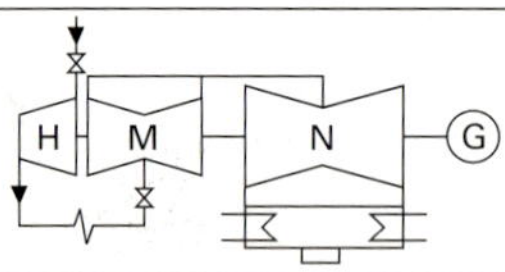	3,000 and 3,600	100 to 1,200	300	600/600
SN Series with reheat	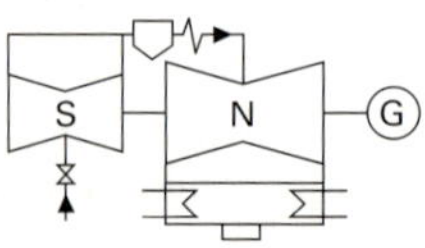	3,000 3,600 and 1,500 1,800	300 to 1,600	70	285

Figure 2.1.1.0 Steam Turbines for Power Plants

2.1 Steam Turbine and Turbogenerator Series

2.1.1 Steam Turbines

The manufacturing range of steam turbines is shown in Figure 2.1.1.0. The steam turbines are designed for use in combined-cycle power plants for dual and triple-pressure cycles with steam reheating, for use in fossil-fired power plants with steam reheating and for use in nuclear power plants.

Steam turbines for combined-cycle power plants with a dual-pressure cycle are of single-casing design as long as the cooling conditions permit (series E). The two-casing design is used for higher outputs featuring a separate HP and double-flow LP casing(s) (series DN). Steam turbines for the triple-pressure cycle with reheat steam are designed with a separate HP casing and a combined IP/LP casing (series HE). For higher outputs the compact design is used consisting of a combined HP/IP casing and a separate double-flow LP casing(s) (series KN).

For reheat steam turbines of pulverized coal-firing power plants or other energy sources using power plants of a smaller rating, there is a choice between the series HE and KN depending on the cooling water temperature and type of cooling. High-rating reheat steam turbines are implemented with a separate HP and IP casing and double-flow LP casings (series HMN).

Steam turbines for nuclear power plants (pressurized and boiling water reactors) are designed with a double-flow HP casing and double-flow LP casings (series SN).

Steam turbine series have been developed by Siemens/KWU for use in the above-mentioned power generating facilities which, as a result of continuous

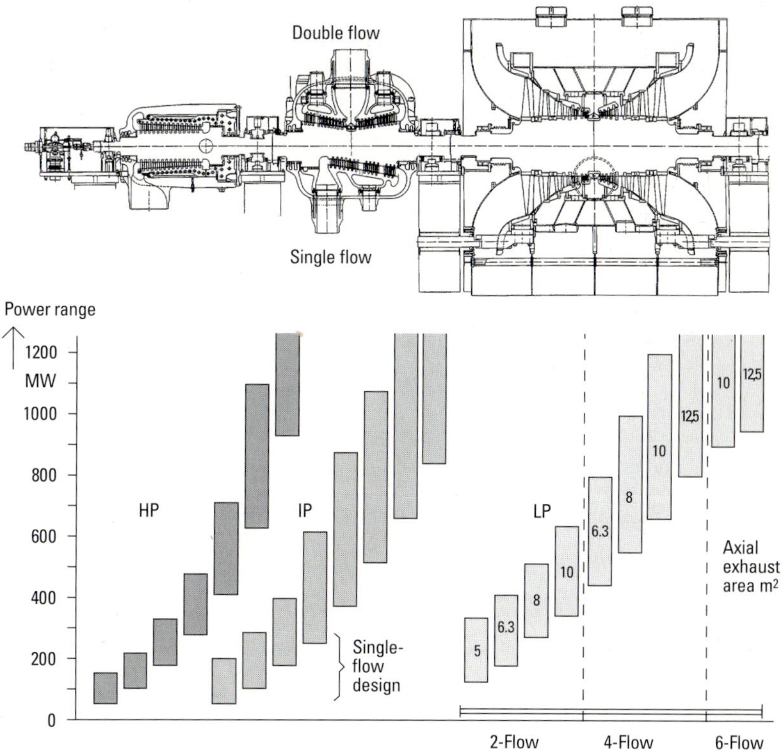

Figure 2.1.1.1 Module Architecture for Reheat Steam Turbines (50 Hz)

evolution, possess a high technical standard. Figure 2.1.1.1 represents as an example the reheat steam turbine (series HMN) for large fossil-fired power plants. One major objective in steam turbine design is highly efficient energy conversion. For large units, separate turbine casings are provided for the individual expansion sections, i.e. the implementation of separate HP/IP and LP casings.

The individual turbine sections can be combined according to the given thermodynamic design conditions and design mass flow. This design series covers the entire spectrum of steam conditions up to ultracritical steam conditions. The structural dimensions of the HP and IP turbine sections are fixed: only the blading is adapted to the given application (steam conditions and rating). The LP sections are standardized turbines, i.e. the dimensions and blades are specific to each model. An optimal grading of the individual modules guarantees the application for the wide range of output and steam parameters. The auxiliary systems of the steam turbine and the I&C systems are also being standardized. The advantage of this concept lies in the large feedback of experience concerning the service reliability of these components, ultimately leading to a high degree of turbine availability. The series-based design concept has been highly successful; a wide range of positive experience with a high number of applications has been gained in operating these turbines.

Table 2.1.2.0 Turbogenerators, Cooling Systems and Rating Ranges

<table>
<tr><th rowspan="2">Type</th><th colspan="2">Stator windings</th><th colspan="2">Rotor windings</th><th rowspan="2">Rating range (typical)</th><th rowspan="2">Main field of application</th></tr>
<tr><th>Coolant</th><th>Cooling method</th><th>Coolant</th><th>Cooling method</th></tr>
<tr><td>TLII</td><td>Air</td><td>Indirect</td><td>Air</td><td>Indirect</td><td>10 to 50 MVA</td><td>Industrial power plants</td></tr>
<tr><td>TLRI</td><td>Air</td><td>Indirect</td><td>Air</td><td>Radial</td><td>30 to 300 MVA</td><td rowspan="2">GUD combined-cycle power plants and coal-fired power plants</td></tr>
<tr><td>THRI</td><td>Hydrogen</td><td>Indirect</td><td>Hydrogen</td><td>Radial</td><td>180 to 430 MVA</td></tr>
<tr><td>THDD</td><td>Hydrogen</td><td>Axial</td><td>Hydrogen</td><td>Axial</td><td>400 to 1000 MVA</td><td>Large coal-fired units</td></tr>
<tr><td>THDF</td><td>Water</td><td>Axial</td><td>Hydrogen</td><td>Axial</td><td>550 to 1500 MVA</td><td>Very large coal-fired units</td></tr>
<tr><td>THDF</td><td>Water</td><td>Axial</td><td>Hydrogen</td><td>Axial</td><td>550 to 1900 MVA</td><td rowspan="2">Nuclear power plants (4-pole generator)</td></tr>
<tr><td>THFF</td><td>Water</td><td>Axial</td><td>Water</td><td>Axial</td><td>1000 to 2200 MVA</td></tr>
</table>

Axial: direct conductor cooling by axial flow through ducts in the conductor.
Radial: direct conductor cooling by radial flow of coolant through cooling ducts perpendicular to conductor.

2.1.2 Turbogenerators

With the individual types of turbogenerators (Table 2.1.2.0) [2], Siemens/ KWU covers a power range from less than 10 to 2,200 MVA. The combination of coolant and method of cooling used for the stator and rotor windings characterize these series. With increases in generator rating, cooling has to be intensified. Whereas air is the principal coolant used in the range up to around 300 MVA, water for direct conductor cooling and hydrogen are also used for high rating machines. In generators with direct water-cooled windings, hydrogen is required to cool the stator core, as is the case with direct, hydrogen-cooled windings. The type of coolant and cooling method dictate the design features of the stator and rotor slots.

2.2 Steam Turbine Generator Sets

The series developed or each type of steam turbine and generator, cover the entire range of applications for power plants operating at a nominal frequency of 50 and 60 Hz for the entire load range, the design parameters and process engineering aspects. The use of proven components and modules ensures high availability, cost-effectiveness and good serviceability. In thermal power plants linked with public and urban energy supply systems, the turbine generator sets are used primarily for power generation. Our modular turbine generator set system is also used in the most varied applications to supply heat for heating purposes and process heat or to recover heat to drive prime movers. For mechanical drive applications and small capacity generators a separate series has been developed.

The turbine generator units are designed for saftey, high availability, long service life and cost-effectiveness. The I&C systems include all equipment of the power plant and turbine generators that are required for operation, automation, monitoring of the processes and also the archiving of process events.

2.2.1 Rating Turbine Generator Series

The development of power plants in the Federal Republic of Germany has always been aimed at the economic aspects of power generation [1]. Upgrading of turbine generator ratings involved a cost degression and an improvement in efficiency. After a phase of increasing the unit size which was particularly pronounced in the case of nuclear power plants because of their high investment costs, the high fuel costs of fossil-fired power plants in addition to the high investment cost for measures in connection with environmental protection resulted in the development of plants of high unit rating with a high energy utilization by the implementation of critical main steam conditions and complex heat cycles. Figure 2.2.1.0 exemplifies the development of the performance data for turbine generators of fossil-fired

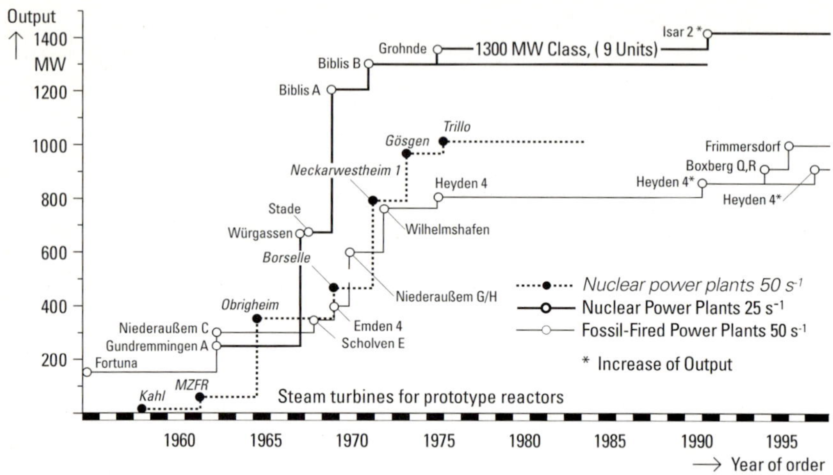

Figure 2.2.1.0 Siemens/KWU Steam Turbine Generator Sets, Increase of Output

power plants and nuclear power plants in Germany which use turbine generators supplied by Siemens/KWU:

- Large turbine generators for fossil-fired power plants and nuclear power plants were built to operate at full speed ($50s^{-1}$) and turbine generators for nuclear power plants particularly in the upper output range had to be designed to operate at half speed ($25s^{-1}$).
- Both designs have proven themselves.
- The cross-compound turbine generator configuration occasionally used, predominantly in the USA, was not used in Germany and in recent years hardly at all in the USA either.

At the present, the rating limit of full-speed tandem-compound turbine generators is governed by the following criteria:

▷ Use of a maximum of three double-flow LP turbines in order to limit axial expansion (efficiency) and deformations of the foundation deck (operational aspects),

▷ Limitation by the axial exhaust area of the LP turbine.

At present, steel blades of 1,150 mm in length (50 Hz frequency) are available from Siemens/KWU, which allow for a nominal axial exhaust area of 12.5 m^2. Titanium blades are being developed which allow a nominal axial exhaust area of 16 m^2. Terminal output of two-pole generators with water-cooled stator windings and gas-cooled rotor has been designed up to about 1,500 MVA. The output limit for four-pole generators with water-cooled stator and gas-cooled rotor windings lies at about 1,900 MVA; the four-pole generator with water-cooled stator and rotor windings has an output limit at about 2,200 MVA.

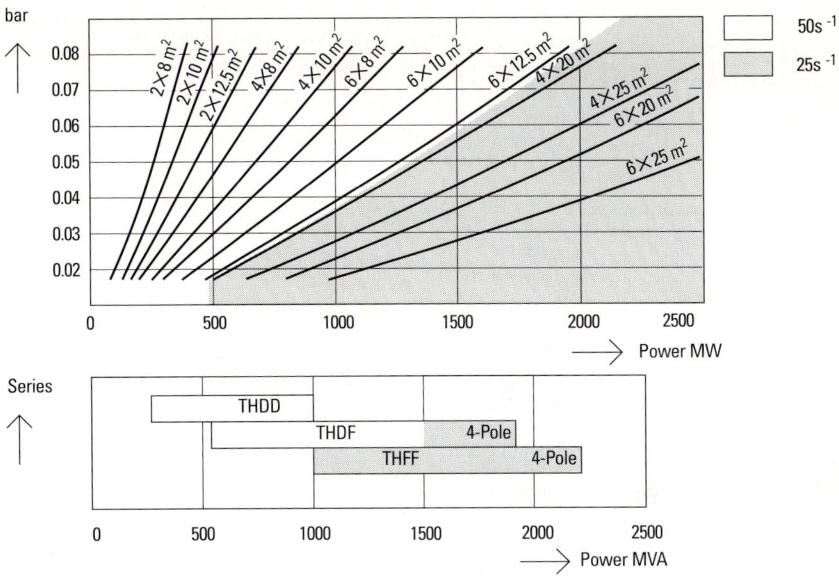

Figure 2.2.1.1 Operating Ranges of Reheat Steam Turbines and Generators

Figure 2.2.1.1 gives an overview of the applications of tandem-compound turbine generators. Increases in turbine generator output are linked to efficiency increases which declines, however, as output increases. At high outputs (1,000 MW class) the efficiency advantage of half-speed turbine generators in terms of heat rate is not the decisive criterion. In the case of saturated steam turbines for light-water reactors, the application limits of the full-speed steam turbine were reached sooner than with the superheated steam turbine because of the smaller isentropic heat drop. If the criteria for the implementation of tandem-compound, full-speed steam turbine generators are not fulfilled, a different design must be chosen. The possible designs are depicted in Figure 2.2.1.2. Investigations of the economic features of the implementation of the individual variants show:

- Tandem-compound half-speed turbine generators of large size have a somewhat better heat rate in the upper output range than tandem-compound, full-speed turbine generators, particularly at low backpressure.
- Compared with tandem-compound full-speed turbine generators, cross-compound turbine generators require higher total costs for the turbine generator, the turbine building and the associated systems and have a slightly lower efficiency than the tandem-compound half-speed turbine generator.

Type of Steam Turbine Generator	Arrangement
Tandem compound (50s $^{-1}$)	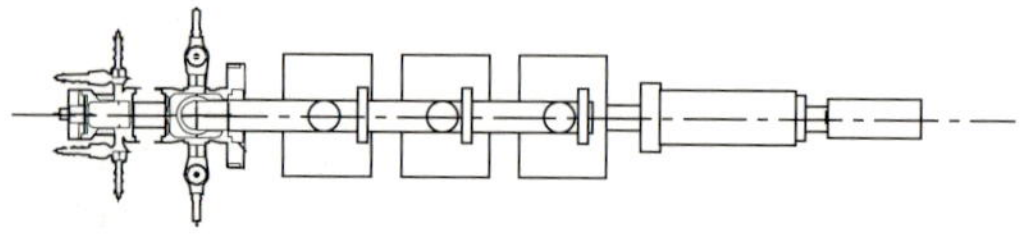
Cross compound (50/50 s $^{-1}$)	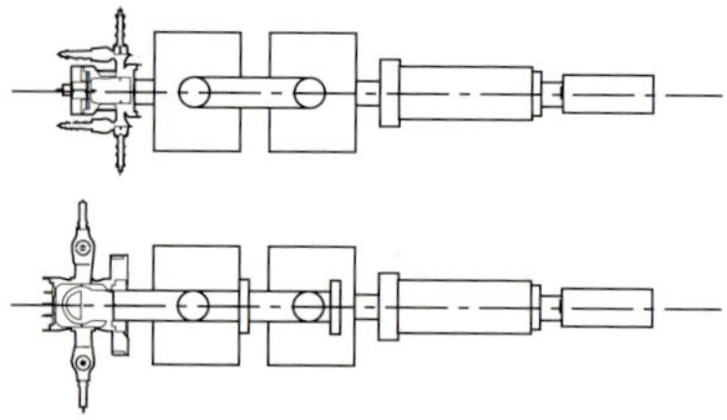
Cross compound (50/25 s $^{-1}$)	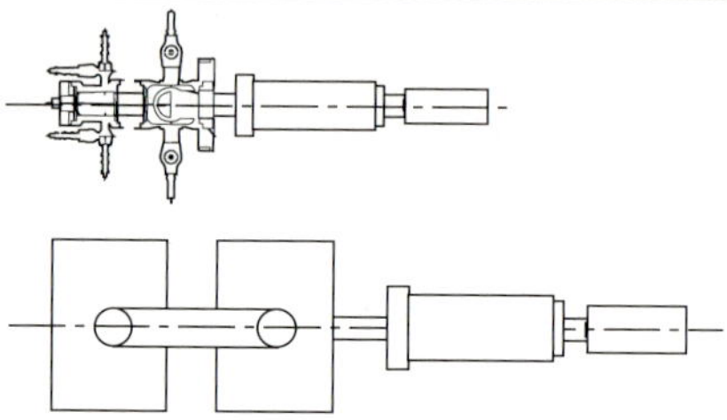
Tandem compound (25 s $^{-1}$)	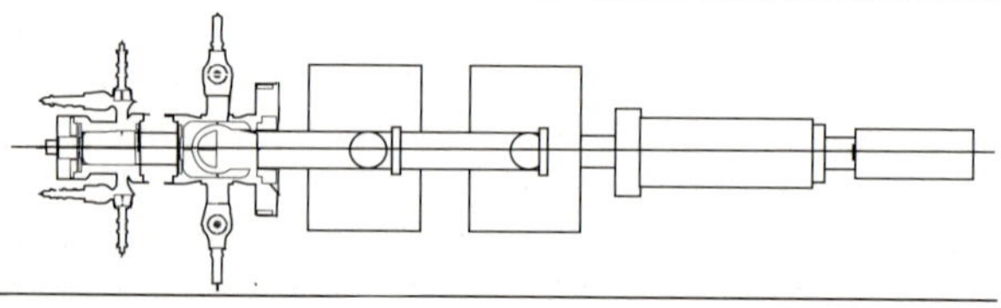

Figure 2.2.1.2 Comparison of Various Types of Reheat Steam Turbine Generator Sets

- In the case of the half-speed turbine generators, equipment weights are high, particularly those of the LP turbines and the generator. Consequently, run-up times to operating speed are somewhat longer for the cold and semi-warm start compared to those of full-speed steam turbines because of the larger dimensions of the half-speed HP and IP turbine rotors and the turbine casings.

On the basis of the present state-of-the-art and the operating experience gathered to date with full-speed tandem-compound steam turbine generator units, this type can be used for high unit ratings with increased steam parameters. The maximum output of full-speed steam turbines continually increased in the past. Technical progress made in the calculation of complex flow dynamics and use of new advanced steels and titanium will permit further increases in maximum output in the future.

At low backpressures the rating of these turbine generators is limited with today's LP turbines and at high backpressure it is dependent upon the generator with water-cooled stator and gas-cooled rotor windings. Cross-compound turbine generators do not have any advantage over half-speed turbine generators. The former were used before sufficiently large last-stage blades for full-speed LP turbines or generators with a correspondingly high rating were available. In the high output range, the choice today is generally between full-speed and half-speed turbine generators. For large capacity units the design selected depends on the design parameters, the heat rate weighting factor and the envisaged operating mode. Siemens/KWU has gathered extensive operating experience with high output half-speed and full-speed tandem-compound turbine generators.

2.3 Operating Mode of Steam Turbines

The steam turbine series available from Siemens/KWU for implementation in fossil-fired power plants are shown in Figure 2.1.1.0. With these steam turbines it is possible to fulfill the requirements relevant to the tasks and load management for the various types of power plants. Siemens/KWU steam turbines have the following essential operating modes:

- Constant pressure with nozzle group control based on partial-arc admission,
- Full-arc admission with variable pressure operation.

Historically, small turbines were built as partial-arc admission turbines with control stages. However, we also provide control stage designs for any requested application, even for large units. In order to achieve high efficiency, especially with larger steam turbines and the application of higher steam pressures, these units were designed for full-arc admission and to operate with variable pressure. The space made available by eliminating the control stage allows the use of additional stages with full-arc admission. This yields higher efficiency at rated load and over the upper load range of the turbine generator unit than that achieved by a design using a control stage. Figure 2.3.0 shows the above-mentioned different types of HP steam turbines. If the different types of design are compared, with this comparison, essentially valid for all steam turbine manufacturers, the following becomes apparent:

The control stage design requires more space than the full-arc admission design. Turbines without a control stage can therefore accommodate, as a rule, up to three additional stages in the blading section of the HP turbine. Since the efficiency of a control stage, even when the steam is admitted to all nozzle segments with fully opened control valves, is lower than that of several full-arc admission stages, with correspondingly smaller isentropic heat drop and associated lower flow velocities in the blading, a turbine with control stage is less efficient than a full-arc admission turbine.

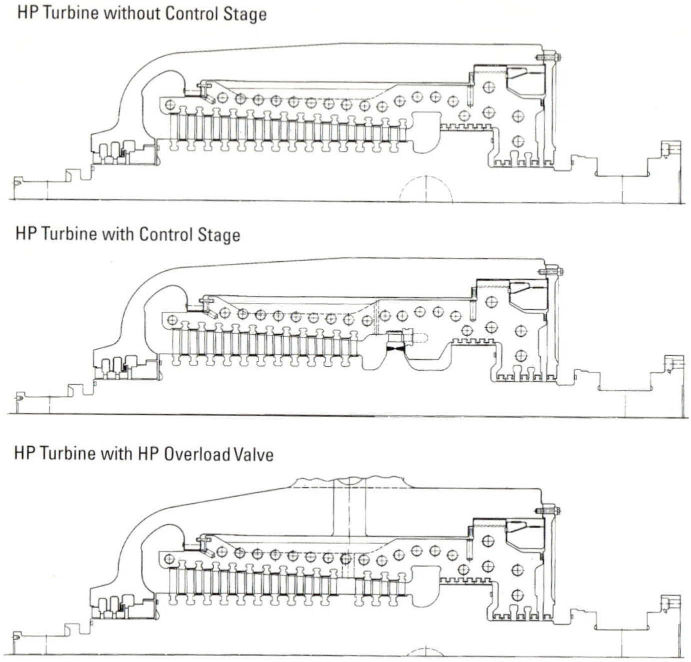

Figure 2.3.0 Different Admission Types of a HP Steam Turbine

For load-cycling turbines, the full-arc admission design drastically reduces thermal stresses because of small load-dependent temperature changes within the full-arc admission. For HP turbines the stresses are roughly half of those in turbines with control stages. In addition to all these operating modes, variable pressure operation results in the lowest thermal stressing of the turbine components under non-steady-state conditions. Due to expansion of the steam in the control stage during fixed pressure operation, the steam temperature changes during transient operation are considerably greater than during variable pressure operation.

The resultant temperature differences in the HP turbine components are correspondingly smaller, which results in lower thermal stresses, or in faster load change rates or shorter start-up times at variable pressure operation. The smaller temperature gradients in the turbine which necessarily occur in the event of transients, result in a longer service life of the steam turbine for the suggested mode of operation.

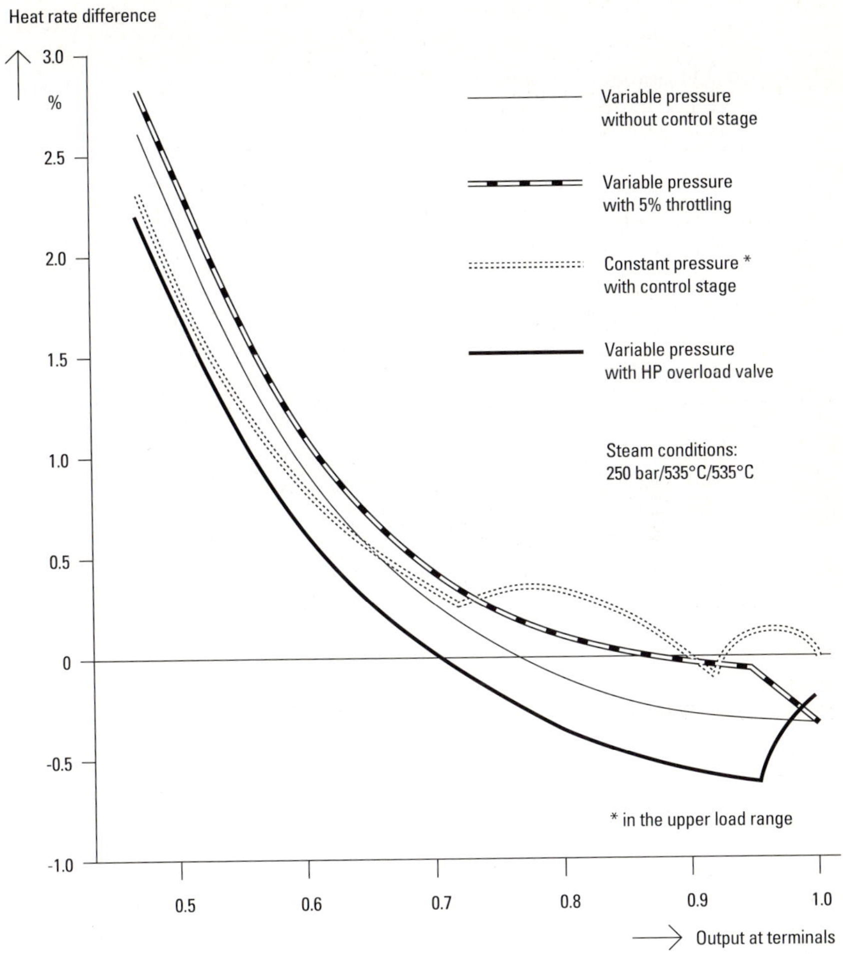

Figure 2.3.1 Comparison of Different Operating Modes of Reheat Steam Turbines

Another advantage of the simple design of the turbine without a control stage is that the high loadings of the oscillating steam forces which occur on partial-arc admission especially in the part-load range do not arise. This advantage should be mentioned since this operating mode can entail negative impacts on reliability. The comparison of the different operating modes with regard to efficiency is depicted in Figure 2.3.1.

Even 5% throttling of the control valves to provide a spinning reserve of 2.5 % of rated load still yields a better heat rate than that achieved with a control stage in the upper load range. In addition to the above-defined methods of initial steam admissions Siemens/KWU proposes the following concept for a large spinning reserve:

- Full-arc admission with an additional overload valve leading steam in a downstream stage of the HP cylinder.

This type is particularly advantageous for steam turbines with large spinning reserves and supercritical steam pressures. This design yields a somewhat higher efficiency than turbines having a control stage at rated output. With the load condition for this example, the efficiency increases considerably in the lower load range without overload valve operation. Other possibilities for realizing a concept with spinning reserves are the condensate stop and/or the deactivation of heat exchangers in connection with full-arc admission.

2.3.1 Definition of the Load Condition for Steam Turbines with Control Stage

The respective load conditions (constant pressure operation) for the different types of admissions are defined as follows:

Rated Output

This is the nominal generator output of the steam turbine (also called 100% load).

Rated Output at Capability Conditions

Nominal (100%) output, but under unfavorable conditions; compared to normal operation increased cooling water temperature (33°C) or increased condenser pressure (0.118 bar) with at the same time preheating of 3% make-up water through the regenerative-preheater.

Maximum Continuous Rating (MCR)

Generator output calculated with the same main steam mass flow as for rated output at capability conditions, but with normal cooling water temperature and zero make-up water.

Valves Wide Open (VWO)

At valves wide open the main steam mass flow is expected to be 3 to 5% higher than at rated output under capability conditions. Compared to rated output at rated conditions, main steam mass flow and generator output is up to 10% higher.

Valves Wide Open/Overpressure

Valves wide open with an increased main steam pressure (pressure increase 5% as compared to the design). As a result, an increase of the main steam mass flow and generator output takes place.

2.3.2 Definition of the Load Condition for Steam Turbines with Full-Arc Admission

The definitions for steam turbine design points using variable pressure operation without control stage are as follows:

Best Point Design (Rated Output)

The turbine is operated at design pressure with the control valves fully open. The condenser pressure corresponds to the condenser design pressure. The swallowing capacity reserve is compensated for by adapting the main steam pressure such that the valves remain fully open.

Overload Point

Maximum overload is reached with the maximum permissible main steam pressure and with the control valves fully open. Main steam pressure is usually 5% higher than at rated output. Any condenser pressure is permissible here.

Maximum Output (Design Capacity)

Maximum steam turbine output is reached with the lowest possible condenser pressure (with choking) and with the maximum permissible main steam pressure. The turbine has been designed for this maximum possible output. Conforming to IEC45-1, the unit may be operated for the specified service life.

Part-Load Points in Variable Pressure Operation

Part-load output can be set in variable pressure operation with the existing condenser pressure and with the valves fully open. The main steam pressure is adapted accordingly to reach this output.

Constant Pressure Operation in the Lower Load Range

Constant pressure operation is adjusted with the control valves throttled. This output is reached by throttling the control valves (constant pressure operation is usually used for low-load operation only, approx. 30%).

Spinning Reserve for Rapid Load Increases

To compensate for a rapid load increase (spinning reserves) for supply demands imposed by the grid, the steam turbine can be operated with the control valves throttled. In consideration of economical plant operation, throttling should not exceed 5 to 10%.

Spinning Reserve for Large Load Increases

In addition to the operation with slightly throttled control valves a number of possibilities, for example "condensate stop" and/or deactivation of preheaters, are available for use in conjunction with a concept or for a turbine without a control stage in variable pressure operation.

Spinning Reserve for Large Rapid Load Increases

A concept with an overload valve implemented for feeding steam to a downstream stage is available at Siemens/KWU for rapid increase of large percentages of the load. This concept allows for part of the main steam flow (usually 10%) to be fed to a downstream stage. The steam turbine is equipped with an additional valve for this.

2.4 Auxiliary Systems of the Steam Turbine Generator Set

Auxiliary systems are required to operate turbine generators. These include the systems for oil supply of the turbine generator, draining of the turbine during start-up, etc. The most important systems are shown in Figure 2.4.0. The Figure shows the arrangement of a 600 MW turbine generator set (series HMN). A description of the auxiliary systems and their function is given in Section 4.

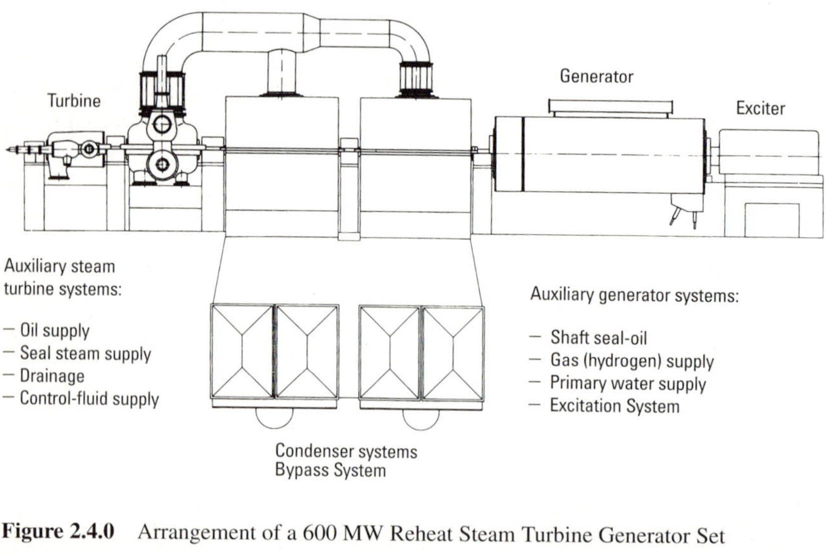

Figure 2.4.0 Arrangement of a 600 MW Reheat Steam Turbine Generator Set

3 Overview of the Process Control and Information System TELEPERM XP

In today's power plants, the complex processes used to obtain a high energy yield and the measures taken to protect the environment place ever increasing demands on process control and the information provided about processes. These demands must be met by the I&C which must also be ergonomically designed for easy operation. Now, more than ever, the I&C equipment in power plants is judged by its ability to improve the cost-effectiveness of power plants. To fulfill this task Siemens has developed the modern digital process control system called TELEPERM® XP for use in power plants of various types and sizes.

TELEPERM stands for the continuous development on the basis of previous I&C systems making use of the experience accumulated. X stands for the use of standard products for both hardware and software. International standards such as UNIX and Windows are used, especially in the area of "operation and monitoring". P stands for Power and shows that TELEPERM XP has been developed especially for use in power plants. This product is therefore optimally adapted to the special demands of power plant operation [3].

The concept is based on the innovations made in microelectronics, software and communication technology. TELEPERM XP is already prepared for the introduction of field bus technology. A modular performance range ensures cost-effective automation. The individual TELEPERM XP modules complement each other in forming a complete I&C system. And the necessity for process control in every power plant to meet individual requirements is achieved by using the concept of individual adaptability of the modules. An essential feature of the concept is a power plant-oriented, easy-to-understand control structure using transparent process-specific subdivisions and avoiding intercoupling. The concept enables a greater degree of user-friendliness and a clear control room concept. TQM (Total Quality Management) ensures that systems are tailored to the customer's requirements and comprises the individual steps taken to draw up a concept right up to service. This concept has been certified to DIN ISO 9001. The structure of the system and its division into subsystems is shown in Figure 3.0.

3.1 Automation Systems for Process Automation

The automation AS 620 system performs the tasks of the group and individual control level, and acquires measured data and states from the process, executes open-loop and closed-loop control functions and then transmits the

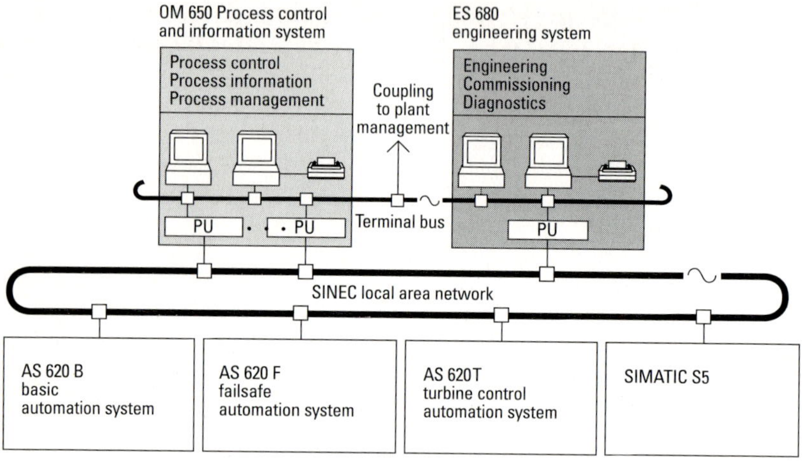

Figure 3.0 OM 650 Process Control and Information System

resulting commands to the process. In accordance with the diversity of the demands made of normal operational measurements, open-loop and closed-loop control, safety-related tasks and fast response to load changes, i.e. in the turbine, the system is build up in three subsystems. As these modules have built-in intelligence, continuation of the control and protection functions is ensured even if the central processor fails.

3.1.1 Subsystems

AS 620 B (Basic)

This system is in use for general automation tasks, such as the protection of an item of auxiliary equipment, open loop control, closed-loop control of the entire unit. This system can be arranged in a centralized or distributed configuration using field buses.

AS 620 F (Fail-Safe)

This system will be used for safety-relevant protection tasks (e.g. at the boiler) and safety controls requiring official approval (e.g. burner controls) with appropriate redundancies.

AS 620 T (Turbine)

This system will be used for the fast closed-loop control of the turbine. The system is based on the latest RISC processor technology of SIMADYN® D and is designed with 1-out-of-2 redundancy to meet the high availability requirements.

SIMATIC S5

SIMATIC® S5 is an automation system developed for the various tasks in power plants. Power plant ancillary systems equipped with SIMATIC S5 can effortlessly be integrated into the unit I&C system. Programmable controllers from the SIMATIC S5 system can be used for I&C tasks in the ancillary systems. Because of its worldwide use, it conforms to national and international standards. The system is also in the care of experienced world-wide operating service teams.

3.2 Bus System SINEC H1
(*SI*emens *NE*twork *C*ommunication *H*igh Range)

This bus system provides a network along which the distributed autonomous components of the process control system, SINEC H1 applies fiber optics and coaxial cables. The bus system fulfills international standards and provides the prerequisites for open communication. The terminal bus in the control room area is of the same product as the plant bus using TCP/IP communication standard. This piece of equipment now makes it possible to access the subcoordinated processing units (PU) from any operator terminal. The terminal bus serves also as a link to other on-site computer, e. g. plant maintenance and management.

3.3 Process Control and Information System OM 650

This process control and information system is the interface between the plant and the personnel in the control room. The plant can be operated, observed and monitored through this extremely ergonomic window to the process. This system also provides all the functions that are necessary for comprehensive logging of events in the process and archiving of data.

The OM 650 operates on PC's on the basis of international standards such as UNIX, X/Windows, OSF/Motif and the database system INFORMIX. The system is modular and has an open structure so that it can be adapted to project-specific requirements and any expansions made at a later stage. The processing units (PU) usually are configured redundantly. The number of monitors can be adapted to the process requirements. The system is operated with a mouse or trackball.

Large-screen displays can be used instead of monitors in the control room to replace conventional panels and the intricate rewiring necessary whenever alterations are made to the process. The large-screen display is not configured separately but provides the same displays as the monitors. It can therefore be easily adapted for individual control design.

3.4 Engineering System ES 680

This engineering system is the central TELEPERM XP engineering and configuration system. The ES 680 is used to configure the AS 620 automation system, the OM 650 process control and information system, the SINEC H1 bus system and the system hardware. The ES 680 has a configuring package for every target system. The ES 680 manages all the configuring data centrally, i.e. the data only have to be entered once. The engineering system software consists of individual modules.

This system uses international standards such as UNIX, X/Windows and OSF/Motif as well as the database system INGRES and INFORMIX. The ES 680 can run on a PC or a workstation, depending on the performance requirement. Function diagrams according to VGB guidelines are used as the basis for the configuration of AS functions and processing functions in the OM 650. The function diagrams are entered in the ES 680 interactively via the function diagram editor. Plant and process diagrams for the process operator control and visualization functions of the OM 650 are also created interactively on-screen using related software modules.

The ES 680 configuring principle is based on consistent forward configuring. Initial configuring and subsequent changes to the configuration are performed in the configuring system, after which the code is automatically generated.

3.5 I&C Concepts

Siemens I&C concepts are geared toward the functions of power plant processes, providing process engineering solutions for the tasks at hand. The new process control philosophy makes allowances for the fact that human capacity to absorb information is limited. This limit is never exceeded to in order to avoid overtaxing human capacity. This process control philosophy was also designed in the knowledge that future power plant control rooms will only be manned by one operator wherever possible. Siemens I&C concepts are based on a function-related structure. A distinction is made between concepts for process engineering function units and concepts that are relevant to the whole plant (process control and information, I&C configuration, reliability, instrument chains, I&C plant structure, diagrams).

The function units are in terms of process engineering functionally complete modules of the power plant process to which I&C concepts for process control/information, closed-loop and open-loop control and measurement are assigned. Siemens I&C concepts are integrated in a clearly structured I&C hierarchy (Figure 3.5.0) which is related to a process engineering functional hierarchy of the power plant process. Its function units are subdivided into more and more subsystems right down to the process.

Large autonomous functions which can be logically linked to one another must replace equipment units, and must organize and combine the existing groups and subgroups, which are based on small units, within large functional areas. A clear, compressed image of the overall process must be created which also includes the plants auxiliary and ancillary systems. Grading or weighting to create “important” or “less important” systems should no longer be necessary.

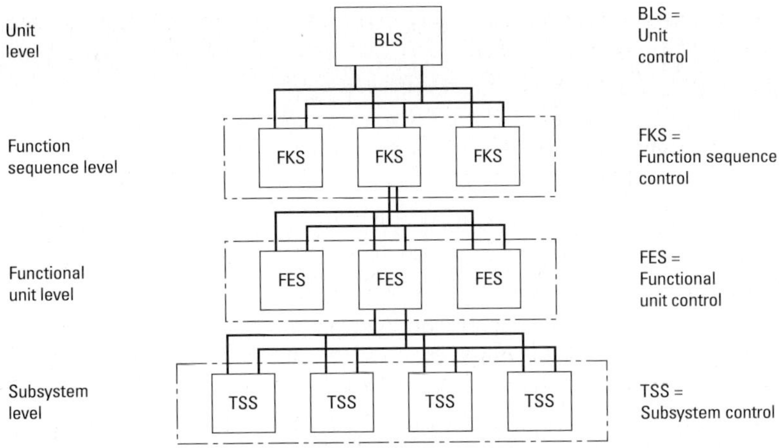

Figure 3.5.0 I&C Hierarchy in Siemens I&C Concepts

If conventional thinking in terms of equipment units, heat exchangers and vessels, auxiliary and ancillary systems and subsystems is rejected, and a power plant process is viewed on a very compressed display of the plant, it is possible to make out process "breakpoints" between which the media and energy are transported. These transport sequences together with the processes that take place at the breakpoints, in a boiler, turbine or pressure vessel for example, all serve the purpose of refining the primary energy, through various intermediate forms, into energy in the form of electrical current. Based on this philosophy, the energy conversion process can be divided into function chains which are defined as transport sequences between process breakpoints.

Equipment units and driven machines, which activate and maintain media flows and transport sequences, move into the background. This division of the power plant process into function chains represents a considerable advance toward the implementation of the cockpit in that it creates an opportunity to condense the information.

Other essential aspects can be read from a very condensed and simplified process image of a power plant. It is clear that function chains can be linked together in such a way that they reflect the operating sequence in all operating phases. These concepts are based on the following principles:

- ▷ Reducing manual intervention and implementation of unit open-loop control in which the control console operator merely has to specify the control commands as well as the setpoints and gradients,
- ▷ Reducing the amount of information by condensing it to a few statements that give a full description of the unit events,
- ▷ Media and signal flow diagrams for the unit overview and function sequences provide condensed information that can be assimilated quickly,

▷ "Positive" annunciations place switching and system positions in the background,

▷ "Negative" annunciations, i. e. fault and incipient fault annunciations, must be greatly reduced by process engineering system correction. The remaining annunciations are weighted according to their significance and are signaled in the displays by a change in color or are simply logged,

▷ Individual control is still possible in principle, e. g. with window management, but manual intervention at the individual control level must remain the exception, even in the event of an accident.

The advantages of Siemens I&C concepts are their simple and modular structure and the ease with which they can be adapted. Specially adapted solutions and high standards of quality improve the cost-effectiveness of power plant operation. Faster fault analysis is made easier by the transparency of Siemens I&C concepts. Past experience based on the same basic concept can be used again. Integrated diagnostics systems that fulfill a variety of different tasks make it possible to detect incipient faults. The modern, ergonomic control console concept can be ideally adapted to customer requirements. Monitors and large-screen displays that can be used in daylight give more scope in the use of light, making it easier to meet the needs of the operating team in an optimum fashion.

3.6 System Test

The quality of the I&C is not only affected by the power of the automation equipment; it also largely depends on the equipment arrangement, the configuration and the interaction of the individual functions. Quality verification can only be provided by an extensive system test. This comprises:

▷ Complete test of the hard and software,

▷ Test of each individual component, where a complete test is not possible for any reason,

▷ Interconnection of all the individual components with the use of suitable simulation equipment for the items of plant equipment concerned.

▷ Preliminary commissioning of the open-loop and closed-loop control equipment (presetting, optimization, taking account of the plant parameters).

▷ Early instruction of commissioning and customer personnel (training),

▷ Acceptance by the customer

The commissioning phase of the plant can be shortened by system tests which has proven successful in turbine generator set I&C over a number of years. The interaction of the I&C components together with a completely simulated plant can be demonstrated using an existing model plant.

4 Steam Turbine Generator Set Monitoring, Control and Protection Systems

4.1 Overview of the Monitoring System

For steam turbine monitoring, those measuring points are provided which give a comprehensive overview of the condition of the steam turbine. Field experience accumulated has shown that an extensive complement of measuring points enables qualified statements to be made on the condition of a plant, increases availability and prevents sequential damage by way of the early detection of incipient faults. Figure 4.1.0 shows the general equipment configuration. The tasks of turbine generator monitoring are:

▷ Speed monitoring during start-up, synchronization, operation and shut-down,

▷ Monitoring of steam temperatures and pressures as process variables, e.g. main steam, turbine admission, extractions, exhaust and condenser pressures and temperatures, casing temperatures,

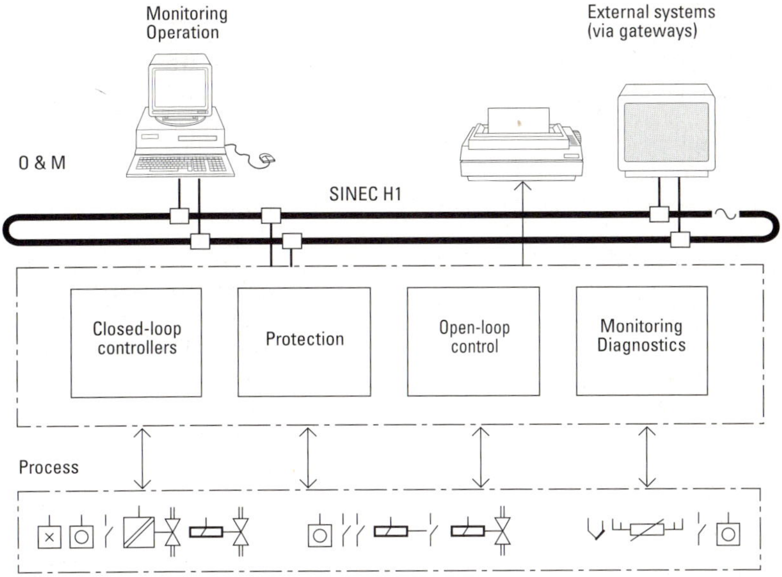

Figure 4.1.0 General Equipment Configuration

- ▷ Monitoring of position of rotor to casings,
- ▷ Monitoring of vibrational behavior,
- ▷ Monitoring of pressures, temperatures and liquid level of oils and control fluids,
- ▷ Start-up of turbine via turbine stress evaluator (margins for speed and power),
- ▷ Monitoring of electrical characteristics in the generator and auxiliary systems,
- ▷ Monitoring of the exciter function,
- ▷ Monitoring of service life expenditure,
- ▷ Analyses and diagnoses of the actual condition of the turbine generator.

The monitoring system comprises:

- ▷ Measured data processing,
- ▷ Data output,
- ▷ Measured data evaluation,
- ▷ Alarm annunciation and display,
- ▷ Data storage.

It provides information and data about the operating condition of the turbine generator. The variables required for this are acquired by measuring appropriate parameters. When set limit values are reached, alarm annunciations accompanied either by audible or visual signals indicate that the measured data in question have exceeded the normal range. In this way, operating personnel can take appropriate measures to restore normal operation of the plant or compensate for a deviation of optimum operating conditions even before a limitation or protection device responds. Analyses, trend calculations, efficiency studies, etc. can be generated using printers, data acquisition systems possibly with long-term storage of whole operational sequences. In the interest of optimum operation and detection of incipient faults, the operational state is monitored under all load conditions.

4.2 Overview of the Control and Protection Systems

The modular instrumentation and control (I&C) concept of power plants is also applied to turbine generators which provides an economic solution to every application what ever concerns the safety and availability requirements and the type of turbine generator involved [4]. The power plant instrumentation and control equipment serves the following functions:

- ▷ Monitoring and Operation,
- ▷ Documentation,
- ▷ Diagnostics,
- ▷ Automatic testing.

Particularly stringent requirements are used in terms of: Reaction time, redundancy, reliability and availability of the I&C systems. Equipment which represents an optimum in terms of performance is closely linked with the task in hand. An essential feature of the subdivision into hierarchic levels with defined interfaces is that a defect in any control device cannot impair the operability of the lower-level equipment. The formation of functional complexes permits clear demarcation lines to be drawn between discrete power plant processes with due allowance for process redundancy considerations. Reliable operation of steam turbine generators places great demands on the automation equipment. The automation concept is organized into the following three levels (Figure 4.2.0):

- ▷ Equipment for normal operation; monitoring equipment (operational closed- and open-loop control, monitoring and diagnostics equipment),
- ▷ Equipment for upset operation; limitation equipment and
- ▷ Protection equipment.

It is the task of the limitation equipment to detect and eliminate abnormal conditions (leaving normal operation) of the turbine without interrupting turbine operation, e.g. limit pressure controllers, limit speed controllers. The protection equipment shuts down either the turbine itself or the items of plant equipment as soon as operating conditions occur that may endanger the turbine or personnel and cannot be held within permissible limits by the limitation equipment, e.g. overspeed protection.

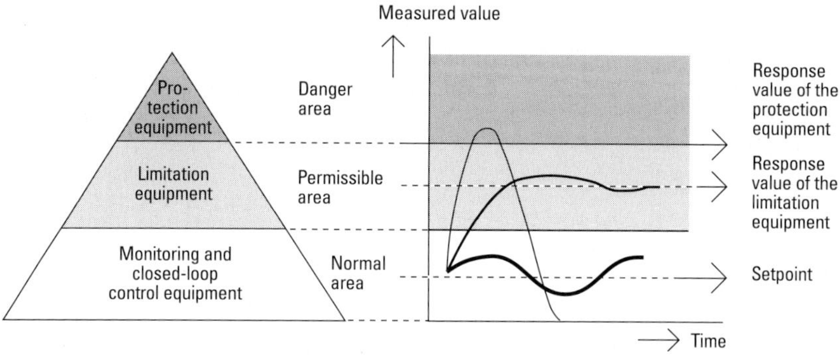

Figure 4.2.0 Monitoring, Limitation and Protection Equipment

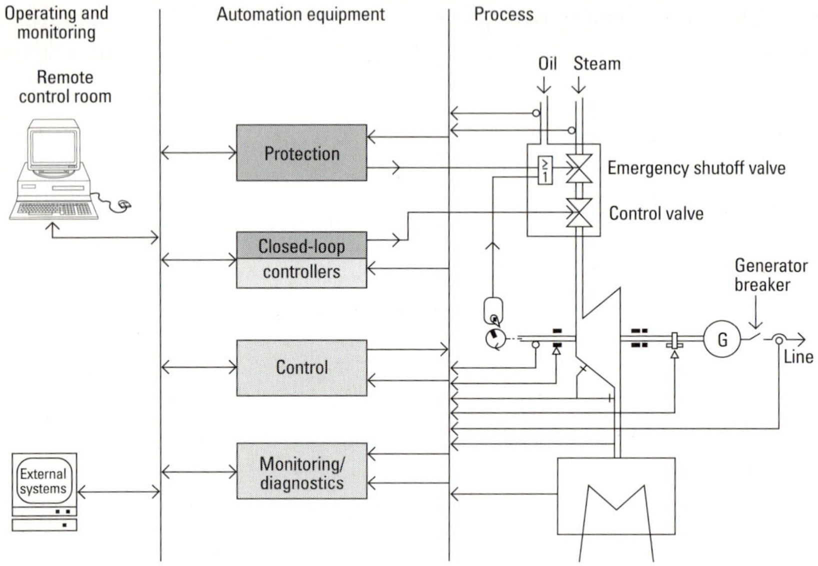

Figure 4.2.1 Basis Structure of Automation System

With its modular structure, the digital I&C equipment offers a tailor-made economic solution adapted to the size of the turbine, plant safety and availability requirements. It has been designed in such a way that single-channel systems for small industrial turbines can be implemented as easily as two-channel and three-channel redundant systems for large plants. Figure 4.2.1 shows the basic structure of automation systems on a turbine generator for protection, closed-loop control, open-loop control, monitoring and diagnostics. Because the I&C configuration used in a plant can be adapted to the plant's needs flexibly and economically, the I&C system is suitable both for new plants and upgrades. The demands of power plant operators can be met with this digital I&C system because of the high standards of quality placed on the hardware and the operating system and the flexibility of the application software. The most important modules of the automation equipment are:

- ▷ Turbine controller,
- ▷ Turbine protection,
- ▷ Generator voltage regulator,
- ▷ Generator protection,
- ▷ Automation system for turbine generator,
- ▷ Turbine generator monitoring and diagnostics.

Only the essential tasks and functions of this equipment are described below.

4.3 Steam Turbine Controller

The tasks of the steam turbine controller are:

- ▷ Starting up and shutting down the turbine,
- ▷ Synchronization with the grid (with the aid of a synchronizing device),
- ▷ Loading the turbine,
- ▷ Influencing the grid regulator,
- ▷ Frequency influence/load shedding while securing auxiliary load,
- ▷ Control of isolated operation,
- ▷ Load ramp duty,
- ▷ Turbine interception during load rejection,
- ▷ Turbine control operation during start-up and in load operation,
- ▷ Pressure stabilization,
- ▷ Limitation of load.

The function of the turbine controller is to supply the steam turbine with the required and allowed flow of steam in all phases of operation. Master controllers which act automatically as required ensure appropriate rates of change for speed or load to match the mode of operation. Setpoint controls, open-loop controls and monitoring functions in the turbine controller prevent the turbine generator unit from reaching inadmissible operating conditions and thereby avoid response of the protection equipment. This forms the basis for high overall availability of the turbine generator unit. The availability of the turbine controller itself is for larger units assured by the redundancy of the control system. The advantages of the turbine controller can be summarized as follows:

- ▷ Stable operation in all phases of operation (start-up, shut-down, parallel operation, island operation),
- ▷ Simple implementation of automation tasks (closed-loop control, open-loop control, monitoring, signalling) through library software functions, functional blocks,
- ▷ Primary frequency control using a linear frequency-to-load characteristic with high sensitivity. A dead-band is provided and can be parameterized as required. The primary frequency component can be activated or deactivated as required from the control room,
- ▷ Safe full-load rejection on disconnection from the grid by limiting transient overspeed excursion safely below the overspeed trip setpoint.

The turbine controller can be interfaced with the unit coordination control system to the extend the control concept requires. Monitoring of thermal stress on important turbine components during start-up, turbine loading and unloading is assured by appropriate interaction with a turbine stress controller or a turbine stress evaluator.

Hardware of Turbine Controller

The turbine controller is realized as a very fast digital system (Siemens SIMADYN®) in a redundant structure (chapter 3.1.1).

4.3.1 Controller Structure

Appropriate to it tasks, the turbine controller makes use of a very fast-responding digital system in a redundant structure. The turbine controller comprises:

- ▷ Master controllers for speed, load and main steam pressure,
- ▷ Subordinate controllers for valve lift,
- ▷ Controller transfer circuit and
- ▷ Start-up device.

Speed and load are controlled by a single combined controller. After synchronization, the speed controller is automatically transformed into a load controller by reconfiguration of its structure with simultaneous provision of a matching feature. If automatic structure reconfiguration is to be inhibited, the operating mode "load operation" with speed controller must then be selected before turbine start-up. This permits operation with pure speed control. If power is already being generated in load control operation, changeover to speed control can be implemented bumpless by selecting "load operation with speed control" in the control room. The output signal from the combined speed/load controller is applied to a MIN logic together with the output signal from the pressure controller. Figure 4.3.1.0 gives an overview of the structure of the electrohydraulic turbine controller.

Speed Control

The function of the speed control is to meet the control requirements of the turbine generator unit during start-up and synchronization with the grid. After

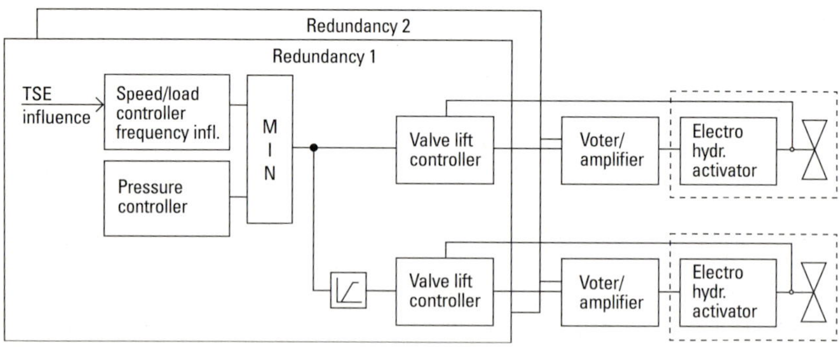

Figure 4.3.1.0 Structure of the Electrohydraulic Turbine Controller

synchronization, the structure of the controller is changed to that of a load controller. After this point, as long as the turbine generator unit is connected to the grid, the speed controller is generally no longer in operation. If necessary, the speed control can also be used for load operation.

The speed control comprises the setpoint adjuster, the setpoint control and the speed controller. The speed setpoint is determined by the setting of the setpoint adjuster. The setpoint adjuster is operated manually from the control room. Further inputs for speed adjustment are provided for an automatic synchronizer as well as for an automatic start-up control. The speed setpoint signal is the input signal to the setpoint controller, which is basically an integrator that determines the rate of speed change. It puts out a modified setpoint signal to the speed controller. The maximum allowable transient is adjusted to meet plant requirements.

The turbine stress controller (TSC) with the turbine stress evaluation (TSE) signal also acts on the setpoint control to limit the speed setpoint. Limitation here is based on the acquisition and evaluation of thermal stresses, measured at representative points of the turbine. This ensures that the turbine is operated within allowable stress limits on the one hand and on the other that the turbine generator unit is run-up with the maximum allowable speed transient.

A distinction is made between an upper and a lower margin for the transients. The upper margin is effective during start-up and loading, and the lower margin is effective for shut-down and unloading. During load operation with the speed controller, the upper and lower margins are converted to give margins that correspond to a load gradient. Depending on the thermal conditions in the turbine, the speed margin may fall to zero or even become negative. This would result in the current speed being held or even reduced, i.e. in temporary closing of the control valves.

The actual speed is measured digitally. Digital speed transmitters are installed at the front-end of the turbine shaft for this purpose. The speed controller features Pl-(proportional-integral) action. This enables no-load speed to be set precisely. As soon as the turbine generator unit is on the grid, the speed controller acts also as a Pl-action load controller. The actual speed is measured digitally in triplicate. For this purpose electrodynamic speed transmitters take their measurements from a gear-like toothed surface section of the rotating turbine shaft.

Load Control

The load control takes over control of the turbine generator unit over the entire load range immediately after the generator has been synchronized with the grid. It consists of the load setpoint adjuster, the load setpoint control and the load controller. The load setpoint is determined by the setting of the setpoint adjuster. The setpoint adjuster of the setpoint can be adjusted by a binary signal from an automatic start-up control. Selection of an external setpoint (grid dispatcher or unit setpoint) permits direct setting of a setpoint through an analog input. The load setpoint is the input signal to the load setpoint control, which sets the maximum allowable load transient.

The implementation and functions of this setpoint control are similar to those of the speed setpoint control. The turbine stress evaluator limits the rate of change of the load setpoint. The TSE signal ensures optimal loading and unloading of the turbine generator unit within the maximum allowable limits. The desired load transient d*P*/d*t* can be set in MW/min on the operation and control system. The output signal from the setpoint control is transmitted to a MIN logic and input to the load controller together with a signal Ps_{max} representing the limit load. The analog input "EXTERNAL LIMIT LOAD" is provided for use in the event, e. g. of steam generator malfunction. The limit load can be reduced automatically through this input.

The load controller comprises two subloops: The first subloop is formed by adding the load setpoint plus the limit frequency dependent component to the output summation point of the Pl-action controller, the second subloop represents the integral load controller proper. The output signal from the load controller is formed by summating the components of the two loops. This structure permits the Pl-action controller on its own to counteract load disturbances caused by the grid. Rapid changes in the load setpoint, on the other hand, are implemented immediately.

The combined speed/load controller always operates as a load controller in normal operation. Provision is made for disabling the load controller, but this is only intended for use in exceptional conditions. The effective load setpoint is always disabled automatically if, in the event of load rejection in response to grid malfunction (e.g., as a consequence, of a 3-pole short-circuit in the grid), actual generator power drops below twice the auxiliary power requirement and the control deviation exceeds about 5% (adjustable), or actual generator output drops rapidly by > 70%, or the generator breaker is opened.

Disabling this setpoint results in a very large control deviation for the load controller. This results in closure of the valves immediately at maximum speed. The controller remains in the load controller configuration. If the actual power level again increases above twice the auxiliary power requirement after a set (adjustable) period, i.e. grid load demand is restored, the load controller automatically restores the operating conditions prevailing before the output signal was disabled. If generator actual power level remains below twice the auxiliary power requirement, the controller is then reconfigured for speed control. Actual generator power is measured by two active power transducers connected to a MAX logic. These transducers are connected to the generator voltage and current transformers.

Pressure Controller

A pressure controller is included in the turbine controller. It serves to assist in maintaining pressure in the steam generator during the limit pressure mode for the turbine. The pressure control allows the operator to manually select two operating modes from the control room; these are the limit pressure mode (boiler follows turbine) and the initial pressure mode (turbine follows boiler). In the limit pressure mode, main steam pressure is maintained at the setpoint by the steam generator pressure controller. Only when the steam pressure drops below an adjustable limit (below nominal operating pressure), for example, in the event of a steam generator malfunction, does the limit pressure controller assist in maintaining pressure by throttling the turbine control valves.

When the main steam pressure in the boiler recovers its normal value, turbine load is automatically returned to the previously set setpoint. To make sure that the pressure control of the turbine controller uses the same setpoint and actual value as the boiler pressure controller, an input for the boiler pressure setpoint and actual value is provided on the turbine controller. Formation of pressure setpoint and actual value is performed for both control systems in the boiler control. In initial pressure mode, main steam pressure in the steam generator is controlled by the turbine pressure controller. The resulting generator power level corresponds to the current steam generator power level (turbine follows boiler). The pressure controller is a Pl-action controller and acts on the turbine control valves via a MIN logic together with the output of the speed/load controller as required by the operating mode selected.

Frequency Influence

The magnitude of the frequency deviation between a reference value (nominal) and the grid frequency is a measure for the ratio between power generation and power export, meaning that under equilibrium conditions (power generated = power exported), grid frequency is equal to nominal frequency. A shortfall in power generation causes grid frequency to fall and excess generation causes this to rise as a function of the magnitude of the power deviation. This correlation is used to counteract frequency deviations in the grid by an appropriate load setpoint correction-signal. This requirement is achieved by the frequency component which is basically a linear function of the grid change in line with the frequency-to-load characteristic and is used to stabilize grid frequency. This means that the load setpoint is raised appropriately in the event of a frequency drop and lowered in the event of a frequency rise.

The two frequency influences are inputs of a MAX logic. The output signal from the MAX logic acts as correction signal on the load setpoint at the summation point of the PI-element. The frequency influence is obtained by comparing turbine speed with the nominal frequency. If required by grid conditions, the frequency influence can be disabled within a dead-band (adjustable) about nominal frequency.

Valve Lift Control

To improve both the stability and the dynamic response of the overall control system, valve lift controllers for the individual HP and IP control valves are implemented as slave controllers for the speed/load and pressure controllers. The valve lift controllers obtain their position setpoint from the MIN logic connecting the speed/load controller with the pressure controller. The actual valve position is measured by a position transmitter at each valve.

Start-Up Device

The start-up device is basically an integrator, which can be stopped by external signals. It provides an analog signal, which is applied to the MIN gate of the controller transfer circuit in order to keep the control valves closed before turbine start-up. At defined positions, it issues limit signals to initiate certain start-up functions, e.g. loading the spring force of the turbine stop valves, and in turn opening the stop valves. This device ensures the correct order of subsequent control functions to be automatically performed during start-up and shut-down. When the turbine is fully in operation, the start-up device is taken out of action by setting its setpoint to its maximum.

4.4 Stop and Control Valves for Main and Reheat Steam

Significant aspects for valve design are safety, reliability and rapid valve positioning. The valves and their actuators are designed based on fail-safe concepts. Material selection is important for frictionless operation. Small numbers of valves are planned where possible, accounting for good accessibility and a minimum of additional connecting pipes.

The stop valves are isolating valves with protective function, being triggered by protective devices to interrupt the steam supply in the event of dangerous operating conditions. For this reason, they are designed for high-speed closing with maximum reliability. The main steam control valves, on the other hand, control the flow of steam to the turbine according to a specified load.

The reheat steam control valves are operative in the lower load range only. Above this load range they are fully open. Their principal function is to throttle the flow of steam from the reheater system in the lower load range and to prevent the turbine from accelerating to trip speed upon load rejection.

The stop valves are single-seat valves with a pilot plug for pressure relief when opening. The stem is sealed from the atmosphere by metal packing and graphite rings located behind the stem leak-off steam extraction. When the valve is fully open, leakage of steam at the valve stem is entirely prevented by back seats on the disc and stem of the valve.

The control valves are equipped with balanced (pipe shaped) plugs which are guided in the valve body covered and protected from lateral impingement. The valve plug has pressure-relief holes to reduce the force required to

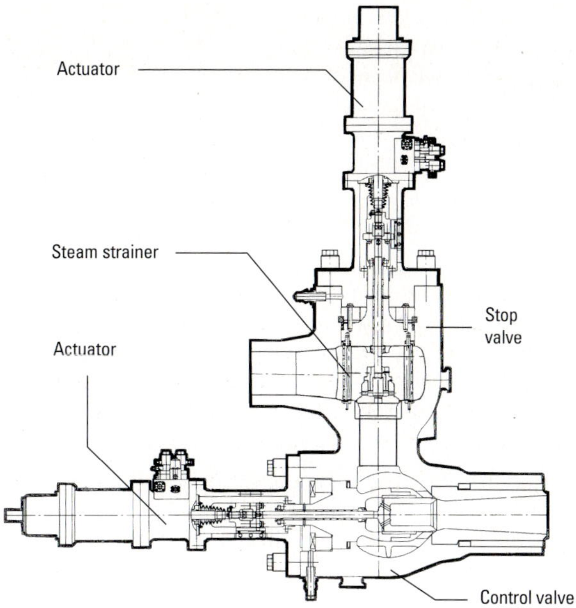

Figure 4.4.0
Combined HP Turbine Stop and Control Valve

operate the valve. The plug of this valve, too, has a back seat which becomes effective when the valve is fully open. Sealing and guiding of the valve stem are the same as in the stop valve. Figure 4.4.0 shows a combined HP stop and control valve. An internal strainer is provided in the stop valve section in order to keep foreign particles away from the admission elements of the turbine and the blading. Fine-mesh wound-strip steam strainers are in use.

The number and size of main steam stop and control valves is a function of the mass flow rate and the type of turbine (nozzle group control or full-arc admission). The number and size of reheat steam valves downstream of the reheater is determined by the mass flow rate. In general combined main steam stop/control valves and combined reheat steam stop/control valves are in use.

4.4.1 Electrohydraulic Actuator for Steam Valves (EHA)

The stop and control valves each have an electrohydraulic actuator with fluid supply by a separate, self-contained high-pressure fluid system. The main features of electrohydraulic actuators for steam valves are:

- ▷ Fail-safe operation due to closing of the stop and control valves by spring force,
- ▷ Accurate positioning by using a high-grade electrohydraulic servo valve and
- ▷ Rapid closing of the stop and control valves (less than 150 ms).

Control valve movement is initiated by the position controller in the event of a control deviation between actual valve position and setpoint. This converter, or so-called electrohydraulic servo valve, is a continuously operating high-grade directional control valve with particularly good steady state and dynamic characteristics and high amplification. A proportional fluid flow is supplied to the hydraulic cylinder as output variable of the electrical input current. The analog feedback signal for valve travel is transmitted by non-contacting position transmitters to the position controller, thereby closing the control loop.

For rapid closing operations resulting from load rejection or turbine trip hydraulic fast valving is employed. For this, two solenoid-operated two-way popped valves are rapidly opened and the two piston chambers via directional control valves and the generous sizing of the two way popped valves allow short closing times and short delay times to be achieved. Figure 4.4.1.0 shows the functional diagram of the electrohydaulic actuator.

The two-way valves are standard cartridge valves with piston guidance. For rapid closure operation the solenoids are deenergized and the control pressure relieved, so that the two-way valve opens and initiate the valve closing operation. The stop valve actuator operates the stop valve in response to binary electric signals such that the valve can only adopt the open or closed

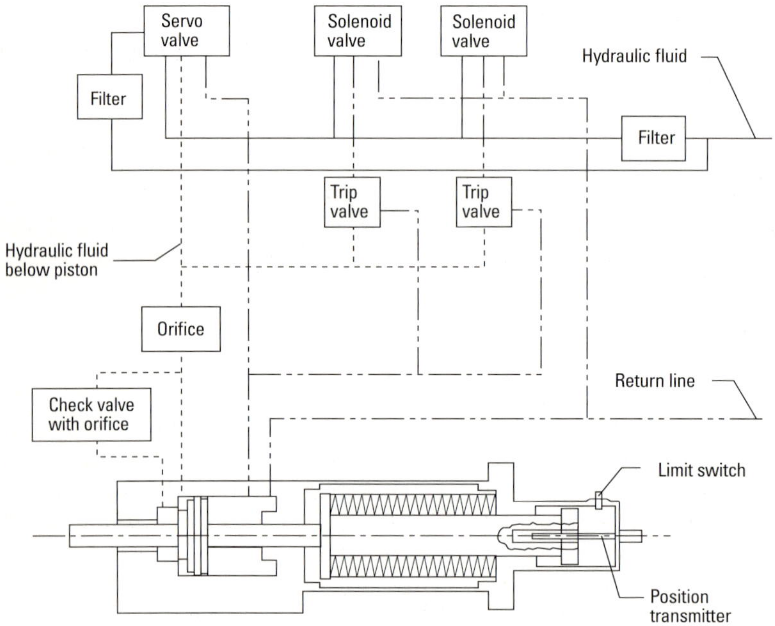

Figure 4.4.1.0 Structure of the Electrohydraulic Actuator for Valves

position. Actuation by pilot-operated two-way valves enables short delays and closing times to be achieved as with the control valve. Instead of the electrohydraulic servo valve a solenoid pilot valve is provided for the stop valve and the limit positions are acquired by non-contacting binary transmitters.

In the cylinder of the actuator cup springs and position transmitters are provided, as well as the control devices which are operated via plates with internal ports. Connection to the central fluid supply unit is provided by a supply fluid line and a return line. The chamber behind the piston is dimensioned so that it can accommodate the fluid quantity, dumped on rapid closure. The return line is therefore of small size. A filter is provided downstream of the fluid line connection to prevent ingress of contamination from the piping system. Power is supplied and signals arc transmitted to the actuator via flexible cabling with plug connectors.

4.4.2 Electrohydraulic Actuator Control Unit

Electrohydraulic Actuators (EHA) are employed in modern turbine installation for operating the turbine stop and control valves. In order to provide the hydraulic power for these valves, a hydraulic supply unit is used. This unit contains the control fluid tank, the pumps, filters and cooling facilities. The electrohydraulic control unit is a device that comprises all functions required for the control and monitoring of the supply unit, the turbine valves, and the testing of the turbine stop and control valves. The unit consists of an electronics cabinet. The electrohydraulic control unit is part of the turbine generator control system; its basic functions are:

- ▷ Control of the fluid supply unit to put the hydraulic system in and out of service and to keep it running safely,
- ▷ Monitoring the actuators and the signals indicating the various operating conditions,
- ▷ Control of the emergency stop valves in accordance with the commands received from superimposed open-loop control devices,
- ▷ Automatic function testing of the turbine stop and control valves by the automatic turbine tester function without interrupting turbine power operation,
- ▷ Automatic function testing of the overspeed protection system without interrupting turbine power operation.

These functions are mainly performed by means of subgroup and subloop controls. The operator intervention is through the control room operating and monitoring system. The electrohydraulic control unit consists of subgroup control units, which automatically run step-by-step routines for various purposes. A further purpose of the subgroup is to put the hydraulic system including the related turbine valves in proper service, maintain its operation, and shut the system down.

4.4.3 Automatic Turbine Valve Tester

Both the emergency stop valves and the control valves are of prime importance in context with turbine protection; it is their function to cut off the steam supply to the turbine immediately, when any protective device has responded. Consequently, it is essential, to test these valves. An automatic turbine valve and subgroup tester is incorporated into the system. The protective devices of a turbine generator unit have to be checked at regular intervals for functional integrity. The automatic turbine tester facilitates regular checking of the protection devices of a turbine generator unit during load operations. It takes the form of a sequential control system.

As it is no longer necessary to shut down the turbine generator unit for the test cycle, such checks can be carried out at more regular intervals. This considerably enhances safety as well as increases availability as a result of early detection of incipient faults. The test sequence is automatically controlled, thus excluding the possibility of human error. The protective devices and the emergency stop and control valves are tested in two subgroups. The automatic turbine tester is operated by means of standardized operation system displays. The test sequence is signalled with the appropriate step and criteria mimics. The automatic turbine tester is switched on via the operation system.

If the corresponding automatic turbine tester has been switched on, the test program can be initiated by the "Start Program" command. In this case all the turbine valves are normally checked in sequence. In the event that the running time for the check is being exceeded as a result of check backs arriving too late or not at all, the current criteria and step signals are stored to facilitate trouble-shooting. A shut-down program is automatically initiated after a fault-free test cycle. In the event that the running time is being exceeded (time-out) or a live trip, the test program is interrupted in midstream and the shut-down program is immediately initiated. The test program can be aborted at any time by the "Shut-down" command. A "Protective Devices" subgroup of the automatic turbine tester is provided and performs the following tests:

- ▷ Testing of the overspeed protection system and
- ▷ Testing of the trip signal paths of the electrical turbine protection devices.

Under normal operating conditions, i.e. when the automatic turbine tester is not operational, the protective devices act directly on the main steam and reheat steam valves via the turbine trip system. If one of the protection criteria is activated, e.g. overspeed protection or condenser protection, the break-current trip bus activated during normal operation of the turbine trip system is tripped. This causes the fast-response solenoid valves of the valve actuators to trip and the turbine valves are closed. A test program is provided for the overspeed protection system in order to test the system and the trip signal path during operation. After the performance of the test, a program available serves to restore the protective devices to their normal operating status.

4.5 Protection System of the Steam Turbine Generator Set

In the event of hazardous operating conditions which cannot be contained by the control equipment (e.g. speed control, limit pressure control units), the turbine generator unit is taken out of operation by the protective devices closing the control and emergency stop valves via the turbine trip system. The protective devices are separated from all other equipment, from the transmitters to the trip loops. The following are groups of protective devices:

- Mandatory turbine protection (absolutely necessary, always included in the scope of supply for the turbine) to protect the turbine against operating condions which could have serious and immediate consequences for persons and property.
- Optional (additional) turbine protection in addition to thc mandatory turbine protection devices.
- Mechanical generator protection, provided essentially to protect the generator when thermal limits are exceeded by insufficient cooling and in the event of leakages.
- Electrical generator protection concept for providing high accuracy, long-term stability and fault detection at an early stage.

The protection system for steam turbine generator sets comprises digital electronic protection circuits equipped with transducers, electronic input/output modules and a high-availability automation system. The protection system monitors essential physical parameters of the turbine generator set against set limits and, if these are found to be exceeded, it first issues a pre-trip alarm, and if a criterion continues to exceed the trip level, it issues turbine trip initiation signals. These signals are transmitted to the individual valve actuators via the turbine trip system. The simplified structure of the functional diagram, e.g. for the steam turbine generator set is given in Figure 4.5.0.

4.5.1 Turbine Protection Criteria

The turbine protection criteria represent the scope of protection required for a steam turbine. They are necessary to ensure safe operation of the turbine and it is not permissible to operate the turbine without these protection circuits.

Mandatory Protection Criteria for Steam Turbine Protection

Overspeed

This protection device prevents overspeed excursions from rising above a specified limit. Overspeed exceeding this limit may cause the rotor to burst due to excessive centrifugal forces, resulting in considerable subsequent damages. Because of the high safety requirements, particularly for this protection, the system is set up in duplicate whereby each system consists of

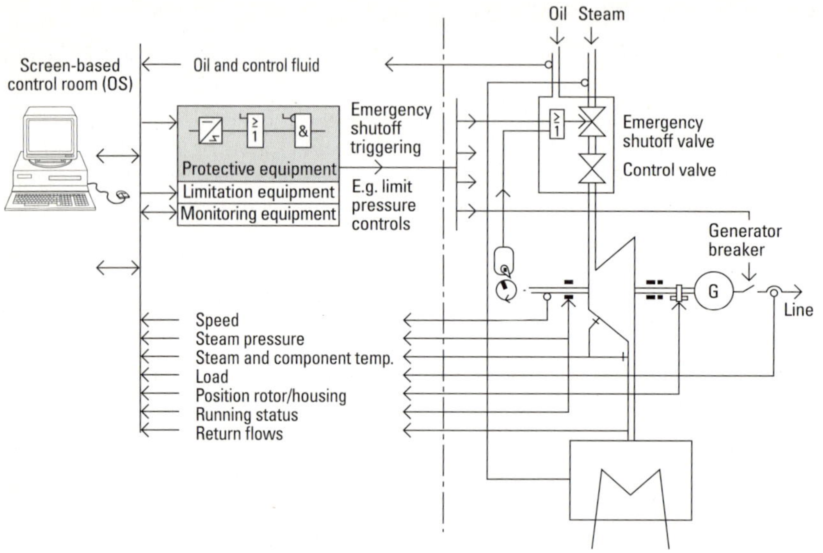

Figure 4.5.0 Function Diagram Protection, Overall Structure, Example

separate 2-out-of-3 deenergize-to-trip circuits. Both systems are tested automatically in regular intervals. The speed is measured by six non-contacting proximity-type sensors.

High Condenser Pressure

Loss of cooling water or excessive air ingress into the condenser causes the condenser vacuum to deteriorate, which in turn will result in an increase of the ventilation losses in the LP turbine. The rise of the LP exhaust hood temperature that follows, will put the turbine at risk. This protection device prevents increase of the backpressure to up to an unacceptable value by first tripping the turbine, and if the backpressure continues to increase, then tripping the LP bypass at a higher backpressure.The backpressure is measured by pressure transducers.

Axial Shaft Displacement (Thrust Bearing Trip)

This protection device is to prevent subsequent damages, which may result from rubbing of rotary blades with stationary, caused by axial displacement of the turbine shaft due to excessive wear of the thrust bearing pads. Turbine trip is initiated if permissible axial displacement in either direction is exceeded. Measurements are taken by independent eddy current sensors.

Low Lubricating Oil Pressure

Dropping lubricating oil pressure often enters in a hazardous situation for the turbine generator bearings. To avoid this, this trip is provided. The trip point setting is such that trip is initiated at the same pressure level, at which the emergency lube oil pump starts automatically. This way control of the situation is maintained and the turbine can safely shut-down. The pressure is measured by means of pressure transducers.

High Bearing Vibration

There are many reasons for rotation-induced vibrations at the turbine generator. These vibrations are transferred to the bearing pedestals and are measured there. If a pre-set limit is exceeded, the turbine will be tripped. Absolute bearing pedestal vibration is measured by piezo-electric accelerometers at each bearing.

Optional Turbine Protection Criteria

In addition to the mandatory turbine protective devices, further protective devices can be provided upon request to protect the steam turbine against various damages.

High Bearing Temperature

The bearing temperature trip circuit protects the thrust and journal bearings of the turbine generator set against excessive temperatures which could cause undue wear of the bearing shells and secondary damage. The temperature is measured with thermocouples in the bearing shell at each bearing including the thrust bearing.

High Exhaust Hood Temperature of the LP Casings

Excessive windage losses put the turbine blading at risk. To protect the turbine blading from excessive temperature, the exhaust hood temperature must not exceed a certain limit. If the set limit is exceeded, the turbine will be tripped. The temperature is measured by thermocouples.

Casing Distortion

The casing distortion trip protects the turbine against damage caused by radial rubbing of the blades or rotor with the casing. Casing distortions are produced if the temperature difference between top and bottom of the HP and IP casings exceed a certain limit, e.g. due to inadequate insulation, steam or condensate backflow from extraction lines. The temperature is measured by thermocouples.

High Exhaust Steam Temperature of the HP Turbine

Under certain operating conditions the exhaust temperature of the HP cylinder may increase to an unacceptable level. The temperature rise is caused due to ventilation of steam in the HP cylinder. The result would be an overheating of the material (blades, rotor and casing) in the HP cylinder exhaust area. The temperature is measured by thermocouples.

High Condensate Level

A turbine trip protects the condenser from excessive condensate levels to avoid consequent floating of the tubes which would deteriorate the performance of the condenser. The condensate level is measured by capacitive level probes.

Low Main Steam Temperature

Boiler disturbance may result in the steam temperature to drop. This will lead to an inadmissible rate of cooling down of the HP turbine, causing inadmissible thermal stress in the turbine components. This condition is counteracted by this protection device by cutting off the steam supply to the turbine. The temperature is measured by thermocouples.

Low Injection Water Flow of LP Bypass

Hot steam being bypassed is not allowed to directly enter the condenser; it has to be cooled down to protect the condenser. For this reason water is injected into the steam flow. The injection water flow is measured and if the flow does no longer correspond with the steam flow, the bypass valve is tripped, i. e. bypassing is discontinued. The flow is measured by a nozzle and differential pressure transducers.

4.5.2 Generator Protection Criteria

Mechanical Generator Protection

It is the task of the mechanical generator protection to initiate a trip when thermal limits are exceeded, and in the event of leakages and other malfunctions in the cooling system. Before the trips are initiated a pre-trip alarm is issued. The following criteria lead to protection trip, inasfar as these are relevant for a specific design:

Cooling-Gas (Air or Hydrogen) Inlet Temperature too High

Following a trouble in the generator cooling system, the cold gas temperature may rise and reach an unacceptable level. Protection against overheating of the components to be cooled is performed. The turbine generator is tripped before the pre-set limit is exceeded. The temperature is measured by resistance temperature detectors.

Hot-Air Temperature in the Excitation Unit too High

To protect the exciter unit from overheating by an excessive air temperature, the trip of the turbine generator is initiated if the pre-set limit is exceeded. The temperature is measured by resistance temperature detectors.

Accumulation of Liquid in Generator, in particular in the Area of Leads, too High

The generator bushing constitutes the lowest point of the generator. Therefore, it is predestined for leaking liquid, which can be both cooling water and oil from the bearings, to collect there. This could end up in terminal and earth faults, when the liquid level rises too high. To prevent the generator from damage, the trip of the turbine generator is initiated. The level is measured by capacitive level probes.

Primary Water Inlet Temperature too High (for THDF and THFF only)

Disturbances in the primary cooling-water circuit can result in a too high cooling water inlet temperature downstream of the cooler. As a result, overheating of the water-cooled components of the generator can occur. If the pre-set limit of the cooling water temperature is reached, the turbine generator is tripped. The temperature will be measured by thermocouples.

Primary Water Flow (Stator Winding, Rotor Winding, Phase Connections) too Low (for THDF and THFF only)

If the primary water flow through the water-cooled stator (and rotor winding for THFF series) is too low an overheating of the water-cooled components takes place and, as a consequence, a destruction of the water-cooled windings occurs. If the pre-set limit is reached, the turbine generator is tripped. The flow is measured by means of nozzles and differential pressure transducers.

Primary Water Tank, Level too Low (for THDF and THFF only)

In case of a too low water level in the primary water storage tank, overheating of the generator components takes place due to the lack of primary water. If the water level of the water storage tank falls short of a pre-set limit, the turbine generator trip is initiated. The level is measured by capacitive level probes.

Electrical Generator Protection

The generator protection system integrated into the unit protection provides the requisite for protecting both the generator and generator leads, and the generator and unit auxiliary transformers. The following are some of the protective functions recommended for generators:

- ▷ Stator ground-fault protection (90 % and 100 %),
- ▷ Rise-in-voltage protection,
- ▷ Overexcitation protection,
- ▷ Underfrequency protection,
- ▷ Differential protection,

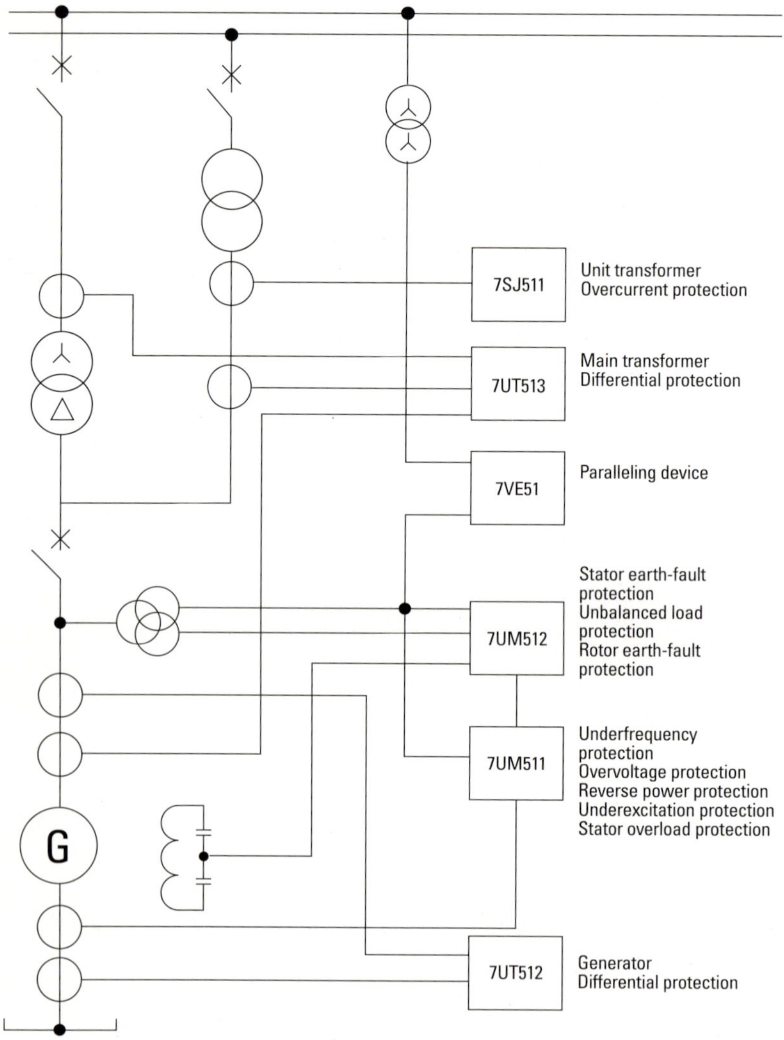

Figure 4.5.2.0 Numerical Machine Protection

- ▷ Rotor ground-fault protection,
- ▷ Load unbalance protection,
- ▷ Distance protection,
- ▷ Underexcitation protection,
- ▷ Reverse-power protection.

The optimal selection of protection characteristics ensures that faults in the generator and the unit itself are detected at an early stage. Depending on the type of fault selective grid disconnection with de-excitation and turbine trip is implemented. Figure 4.5.2.0 gives an overview of a modern numerical protection, as an example.

4.5.3 Fire Protection for the Steam Turbine Generator Set

Fire hazard cannot be ruled out completely, even in the case of steam turbine generator units of a high technical standard. For this reason, a special fire protection concept has been elaborated for the turbine generator unit which should be integrated into the overall plant fire protection concept. In view of the special conditions pertaining in the event of fires and fire hazard, appropriate protection and control measures can be performed manually by actuation of a fire protection switch. Decision-making is facilitated by the status check-back signals from the lubricating oil system monitoring and alarm annunciation systems. The fire protection switch comprises a number of push-buttons connected in parallel, one being provided in the control room and the others at appropriate locations in the turbine building. Actuation of the fire protection switch can initiate the following measures:

- ▷ Turbine trip,
- ▷ Cutting out of the full-load auxiliary oil pumps,
- ▷ Cutting in of the jacking oil pump,
- ▷ Cutting in of the emergency oil pump.

4.5.4 Structure of the Turbine Generator Protection Equipment

The protection equipment for the turbine and generator includes:

- ▷ Measured-data processing,
- ▷ Limit signal formation,
- ▷ Signal gating,
- ▷ Trip signal formation,
- ▷ Protective logics.

The trip criteria (e.g. speed, pressure, bearing temperatures) are processed and than monitored by the limit signal formation. If a dangerous condition occurs, a trip signal (open-circuit or closed-circuit principle) is formed in the signal gating (majority polling circuit depends on redundancy), which shuts down the turbine by actuating the turbine valves. The trip signals are annunciated.

Protection Circuit Configuration

Standard turbine protection circuits consist of a three channel arrangement in 2-out-of-3 configuration. There are some circuits in 2-out-of-2 configuration. They comprise equipment for measured value acquisition, signal conditioning and processing, limit signal formation, trip signal formation and protection signal output. Where applicable pre-trip alarms are also issued before the trip limit has been reached.

4.5.5 Turbine Trip System

The signal processingst of the trip system is a double redundancy system 1-out-of-2 to reach a high degree of safety and availability. The manual actuated turbine trip system is an electrical two-channel system functioning through a 1-out-of-2 relay trip logic. It constitutes the link between the protective system and the turbine "Trip Solenoid Valves". The "Trip Solenoid Valves" operate in a deenergize trip fashion (fail-safe). This will cause the turbine to trip by actuation of the "Remote Trip Solenoid Valves". Turbine trip signals that are derived from the balance of plant, but do not require the generator to be disconnected from the transmission system simultaneously, are wired so, that the generator breaker and field breaker are not tripped directly, but by the "Reverse Power Protection". Also during normal operational shut-downs of the turbine generator the generator main breaker and field breaker should be switched off through the "Reverse Power Protection" after the turbine has been unloaded completely. The turbine trip system receives its signals from the following protection systems:

- ▷ Electronic turbine generator protection system,
- ▷ Manual remote trip from the turbine control panel,
- ▷ Generator protection system,
- ▷ Fire protection system,
- ▷ Collection of turbine trip signals from other equipment.

Response of one protective device always causes both normally energized channels to be deenergized. Turbine trip is initiated if either one of the two channels is deenergized. The trip channel of the turbine trip system is checked out together with the protective devices and the overspeed trips. With this fail-safe arrangement and with the inputs and outputs automatically tested, a maximum in reliability and availability is ensured.

4.6 Service Life Control

As a result of the presumed service life of a steam turbine, the number of starts and load changes are given by the material fatigue. The number of starts is depicted in Figure 4.6.0. The run-up numbers are valid for the steam turbine, any restrictions of other components of the power station are not considered

Number of permissible starts *

	• 1100 starts	with forced cooling during the subsequent shutdown processes
	• 1100 starts	with natural cooling during the subsequent shutdown processes (i.e. without forced cooling by steam temperature reduction/throttling)
Σ ⇒	• 2200 load changes	in fixed-pressure operating mode only
	• Load changes	in variable pressure operating mode without restriction (with almost constant main steam temperature)

The following initial states can be taken as a definition of the terms cold, warm and hot start:

Meaning of the types of start

	Central component temperature to VGB guideline VGB-R105M	Frequency of starts
Cold start	< 160 °C	200
Warm start	160 – 450 °C	400
Hot start	> 450 °C	1600
		Σ 2200

For running the turbine up to rated speed, a temperature of ≥ 20 °C is required for all turbine rotors.

Applies to power plants with conventional steam generators with single reheat and for combined cycles (GUD) with heat-recovery steam generator.

*) using the permissible limits for the thermal stress on the relevant components

Figure 4.6.0 Standard Load Combinations for Steam Turbines

by the rates calculated. Load changes with variable pressure operation on nearly constant main steam and reheat steam temperatures are without significant influence on life time. All start-up procedures and load changes will be controlled by the turbine stress evaluator and as an extension stage, the actual life time expenditure can be calculated by the turbine stress controller.

4.6.1 Start-Up of the Steam Turbine Set (Low-Alloyed Material)

The start-up curves cover the start-up conditions of the steam turbine where speed, load, pressure and temperature vary with time. Under these conditions the parts of the turbine in contact with the steam are subjected to changes in temperature, which result in thermal expansion and in material fatigue due to restrained expansion. To control material fatigue in order to achieve the planed life time the rate and number of temperature changes has to be controlled.

Additionally in the interests of safe operation the axial and radial clearances between the rotating and stationary parts are sufficient at any normal and abnormal conditions, however, relative expansion and factors such as vibration, casing deformation, etc. also require consideration. The temperature changes in the turbine components depend primarily upon the following factors:

▷ Changing steam temperatures during start-up and shut-down,

▷ Changes in turbine load.

As a rule, the turbines are fitted with a turbine stress evaluator which is integrated in the turbine control system. With the stress evaluator the operator can select from different modes to achieve the required start-up times. Figure 4.6.1.0 shows the start-up curves for cold, warm and hot start conditions of a 300 MW reheat steam turbine operating up to 540 °C steam temperatures. The data in the curves should be regarded as an example for the "normal" start-up mode and are valid for the dimensions of the individual turbine type, the given steam conditions of the steam generator and use of the full computational life time. The run-up time begins with the opening of the control valves. Speed and load are increased in line with the curves. However, passage through the critical speed range is to be performed rapidly and

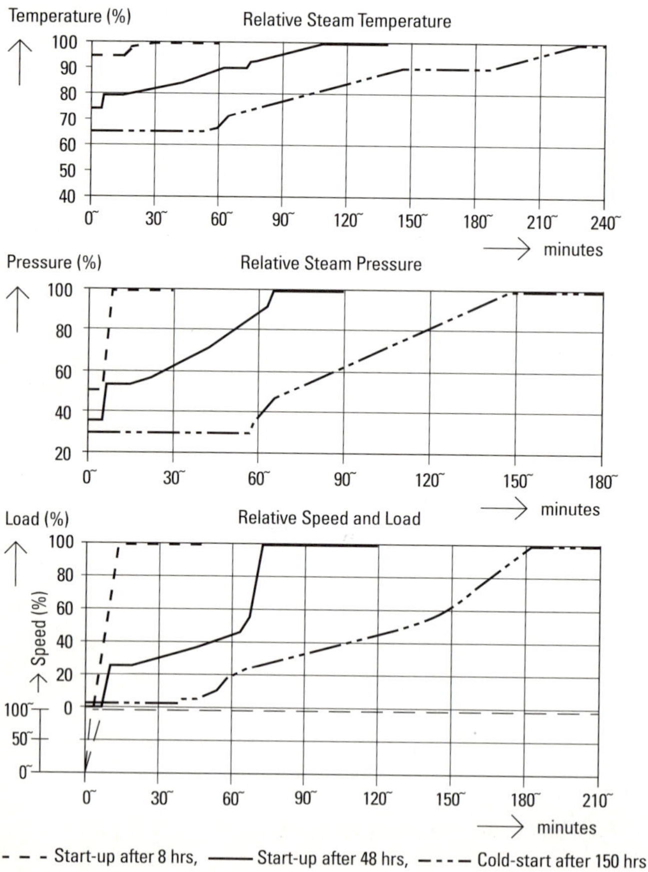

Figure 4.6.1.0
Steam Turbine Start-Up Curves for Various Shut-Down Times

without stopping. The curves give steam pressures and temperatures ahead of the main stop valves as a function of time. Prewarming of the turbine can commence as soon as the criteria for opening of the control valves have been fulfilled:

- The steam upstream of the turbine must be superheated,
- To avoid undesired cooling, the main steam temperature should be higher than the temperature of the turbine components before turbine roll-off.

The control valves are opened by the speed controller, the setpoint of which is set to the prewarming speed, which must be below the barred speed range. Feed forward the minimum temperature margin of the turbine stress evaluator to the setpoint control for the speed controller ensures that the turbine will not be subjccted to excessive thermal stresses. After turbine roll-off, turning gear operation is terminated. The turbine generator unit is run-up to rated speed as soon as the criteria for run-up to speed have been fulfilled:

- The shaft temperature should be at least 20 °C.
- To avoid excessive thermal stress and delays, the main steam temperature must not be too high with respect to the temperature of the HP turbine before turbine run-up to rated speed.
- Before the turbine is run-up to rated speed there must be an adequate temperature margin in the components to permit acceleration of the turbine via the speed controller without any limitation by the speed controller setpoint control to avoid excessive thermal stresses.

The turbine generator unit is run-up to speed by means of the speed controller, with the setpoint control for this rapidly raising the setpoint to rated speed. The speed setpoint is tracked to this target setpoint at a fixed rate of speed change to ensure rapid passage through the barred speed range. If – as a result of the steam mass flow admitted – during acceleration to rated speed any of the temperature margins in the turbine components falls too much in magnitude, a signal acts on this setpoint control to reduce the rate of speed change.

If a stress limit is violated, the monitoring function is activated to reduce the speed setpoint to prewarming speed. The prewarming procedure must then be extended until the temperature condition of the turbine makes it possible to run-up to rated speed without intervention from the stress evaluator. One of the conditions for prewarming and run-up of the turbine to rated speed is that the subloop control for the drains must be in operation. All drains whose criteria for closure are fulfilled in the course of the turbine start-up steps described above must then be closed. Particular attention must be paid to the following operating parameters during turbine run-up to rated speed:

- ▷ Vibration,
- ▷ Bearing metal temperatures,
- ▷ Temperature differences between casing top and bottom halves,
- ▷ Oil temperature downstream of the oil cooler.

If any of these operating parameters exhibit differences from the values previously observed under the same conditions, the reasons must be identified and, if necessary, eliminated. The generator is synchronized after the frequency, voltage and phase angle of the generator have been matched to the frequency and voltage of the power system. Once a certain output level which depends on the plant's own auxiliary load requirement has been reached, the turbine controller changes over automatically from speed control to load control.

4.6.2 Start-Up and Load Changes for High-Alloyed Materials

Components made from high-temperature chromium and austenitic steels require a longer warm-up time than identical components made from ferritic steels [1]. In order to determine the required duration of component warm-up, material properties and design-related notch effects must be considered. The warm-up times of components made from various steels can be compared on the basis of their allowable thermal ramp rates. For components of identical dimensions, this ramp rate is essentially expressed by the equation given in Figure 4.6.2.0, which shows the relation of ramp rates of the IP turbine section shaft of a large-capacity steam turbine of various high-temperature steels and superalloys for high temperature application.

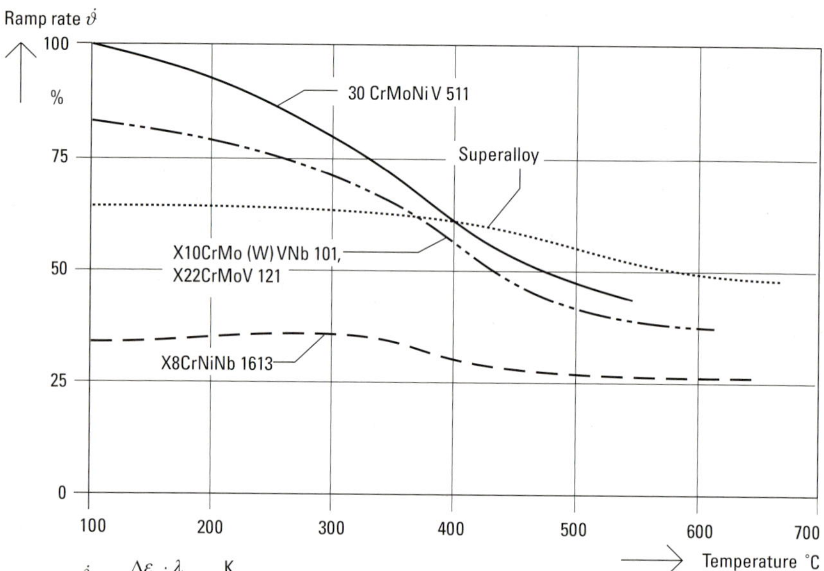

$$\dot{\vartheta} \sim \frac{\Delta\varepsilon \cdot \lambda}{\beta \cdot c} \quad \frac{\text{K}}{\text{min}}$$

$\Delta\varepsilon$ 10^{-3} ‰ Total strain range

β $\frac{1}{\text{K}}$ Coefficient of thermal expansion

λ $\frac{\text{W}}{\text{mK}}$ Thermal conductivity

c $\frac{\text{kJ}}{\text{kg K}}$ Heat capacity

Figure 4.6.2.0 Turbine Rotor Thermal Ramp Rates for Different Materials

Compared to other steels, austenitic steel has the lowest ramp rate due to its lower thermal conductivity and larger linear expansion coefficient, and its warm-up time is therefore the longest. 12% CrMoV steel is closer to ferritic steel with regard to ramp rate and warm-up time. In the upper temperature range the ramp rate of the superalloys is higher than that of the above steels. Tests performed to date by various steel makers can be adopted to large steam turbine components. In the case of start-up from the cold shut-down condition which occurs relatively rarely, the differences between the ramp rates of the austenitic and ferritic materials are considerable. These differences are not so great in the case of load changes in the upper temperature range and during hot starts.

4.6.3 Turbine Stress Controller (TSC)

The turbine stress controller (TSC) is a turbine monitoring device that continuously evaluates relevant turbine components in order to provide necessary information for the turbine to be optimally operated with regard to minimizing life expenditure, but at the same time allowing maximum operating flexibility. It serves the following purposes:

▷ To keep the turbine within acceptable stress limits under all operational conditions by means of the electric turbine controller and

▷ To calculate and record life time expenditure of relevant turbine components.

For this reason, the TSC calculates transient thermal stresses for the relevant turbine components. This monitoring permits smooth operation of the turbine with operator-selectable stress limit settings under all operation con-ditions. There are the three distinctive limit settings to be selected by the operator, one each for:

- Slow,
- Normal,
- Fast.

Making use of these three modes, the operator will be able to achieve an optimal turbine start-up. In practice, this means that, for example, the relatively infrequent turbine cold start, which requires considerable more time, can be performed more rapidly by accepting a somewhat higher accumulated fatigue rate so that ultimately energy is saved. The resulting lower service life can be compensated at appropriate times by low stress operation.

The TSC utilizes temperature differences as a measure of thermal material stress in turbine components, and compares these differences with permissible values, resulting in temperature margins.

Concept of the TSC

The temperature signals are provided by special wall temperature sensors, which are thermocouples. The measurements are taken at characteristic locations of the turbine casings, where high material stresses are expected to occur. Temperature measurements are taken from the temperature of the steam-wetted inside surface of the casing wall and from the centerline temperature of the respective casing wall. Both thermocouples are combined to form the special wall temperature sensor. The thermal behavior of the turbine shaft is simulated electronically by taking one casing temperature measurement at the inside surface that is considered representative for the shaft surface. From this measurement, the mean integral shaft temperature is calculated.

One configurable monitoring channel is used for each turbine component (valves, casings, shafts). The Figure 4.6.3.0 shows the turbine stress controller configuration with influence on setpoint controllers for the turbine control system. In each circuit, temperature differences are derived from the measured temperatures. These differences are compared with the temperature-dependent permissible temperature difference. This comparison yields the respective temperature margins. The relevant upper margin is used for

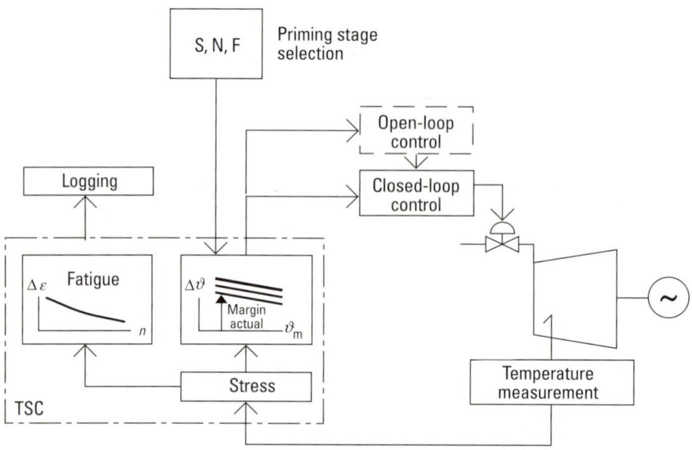

$\Delta\varepsilon$	Fatigue component	$\Delta\vartheta$	Temperature difference
n	Crack initiation life	ϑ_m	Mean component temperature
S, N, F	Slow, normal, fast		

Figure 4.6.3.0 Turbine Stress Controller Configuration

raising the turbine's temperature (increasing speed or power), the relevant lower margin is used for temperature or load reduction. The smallest temperature margin is transmitted to the setpoint controllers of the turbine control system for use in loading or unloading the turbine.

The turbine stress controller checks the transient thermal loads on the turbine during operation. This permits monitored operation even with different permissible loads during all operating phases. During operation with each of the three different load limits, the desired limit is selected by switch command and the occurring fatigue rates of the monitored assembly components are continuously recorded and evaluated. In addition to assembly component monitoring, the turbine stress controller performs accumulated fatigue analyses and long-term data storage for the assembly components to be monitored.

When the steam temperatures in the interior of the turbine change, the metal temperatures in the relevant components follow after a time lag. As a result, free expansion and thermal strains occur. The latter cause material stresses of thermal loads, which lead to corresponding accumulated fatigue to the material. During operation the loads occur cyclically in alternating directions (compressive and tensile stress) due to the heating and cooling in regions of assembly component surfaces exposed to steam and in the shaft axis. The determining factors for the magnitude of the temperature differences in the assembly components and thus for the thermal loads are the time sequences of the steam temperature changes, the heat transfer from steam to metal, and the geometric shape and the material properties of the assembly component.

Calculation of Life Expenditure

Turbine components are designed for the required total operating life of 200,000 h. This is done by using the lifetime computing program created for this purpose. The computational life time incorporates the minimum values of material for life time expenditure calculation which ensures that the planned service life is definitely reached or exceeded. The continuous recording and cumulative totaling of the occurring material fatigue allows transient processes to occur with different start-up rates depending on the requirements (three operation modes). A fatigue analysis is performed for the respective components surface taking the notch geometry and the planned service loading combination into account [5]. As an example, Figure 4.6.3.1 shows the calculation procedure of a turbine shaft. The stresses calculated are combined to form a total stress intensity. The occurring material fatigue values are recorded and the thermal loads are computed. The material fatigue consists of three portions:

- ▷ Accumulated low cycle fatigue,
- ▷ Accumulated creep fatigue and
- ▷ Accumulated stress cycle fatigue (crack growth).

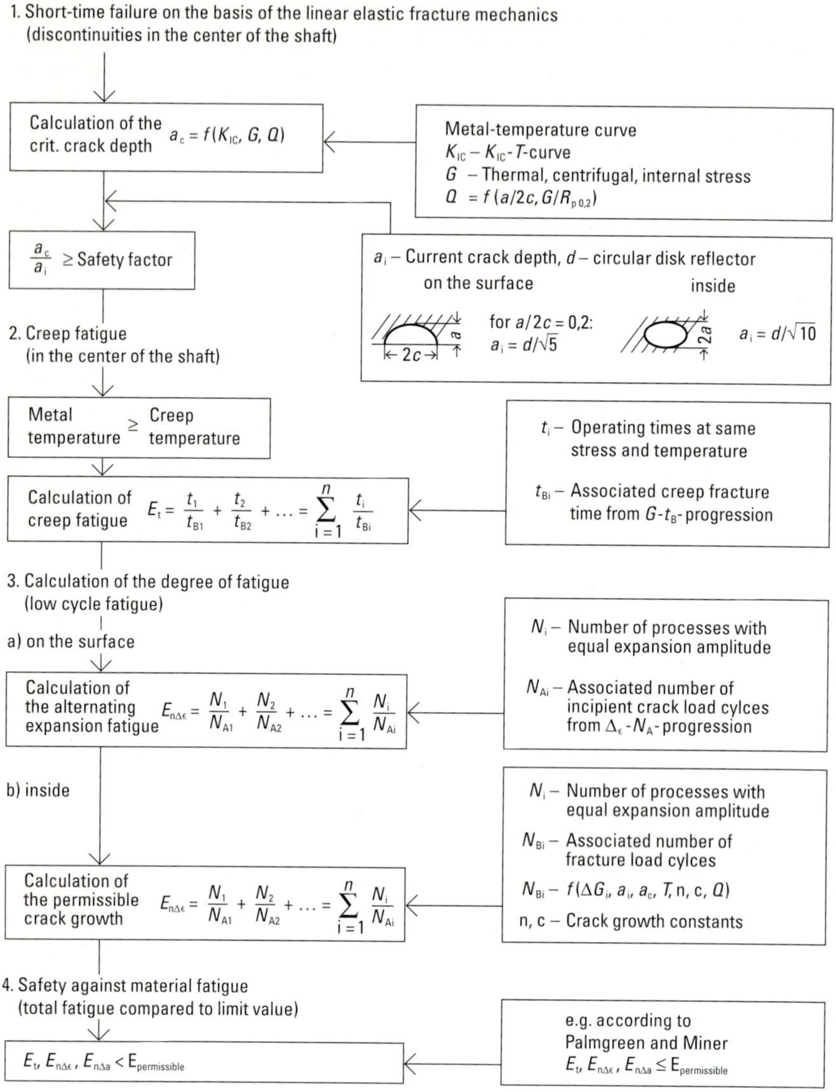

Figure 4.6.3.1 Assessment of Service Life of Turbine Shafts

Accumulated fatigue computations are performed for the relevant components (all measuring points) including accumulated low cycle fatigue and accumulated creep fatigue. For the turbine shafts (HP and IP), in addition, the accumulated stress cycle fatigue (growth of a fictive crack) is also computed.

The respective extent of accumulated fatigue, separated into accumulated fatigue low cycle strain, accumulated creep fatigue and accumulated stress cycle fatigue (crack growth) are displayed separately for each assembly component. During the accumulated fatigue computation long-term storage of the basic data (e.g. temperature, pressure and time interval to the last event) is undertaken for the representative assembly components. These basic data are saved daily on hard disk and monthly on an optical disk. A temperature-pressure matrix providing information on how many hours the assembly component was exposed to the corresponding temperature and pressure is set up for each assembly component.

Accumulated Low Cycle Fatigue

The basis for computation of the low cycle fatigue is the occurrence of temperature differences and resulting values of the strain range of the steel. They occur at the steam side surfaces (casing: inner surface, shaft: outer surface). Different strain range values, occurring successively in time, are defined as an intermediate load cycle and are computed and temporarily stored during a main load cycle. A main load cycle is defined as the time sequence of one positive and one negative strain range value. When the main load cycle is completed, the accumulated fatigue portions of main and intermediate load cycle are added to the total accumulated fatigue of the assembly component. Thereafter the printout contains in addition to other values, information about the surface temperatures of the components, different strain ranges and accumulated fatigue status.

Accumulated Creep Fatigue

The design of assembly components which are exposed to high operating temperatures is generally based on time-dependent strength values (rupture strength). The assembly components are dimensioned for a specified operating temperature and thus for a specified load. The steam pressure and temperature of the turbine section are required as measured data. If operating temperature and load are constant over the operating period, the accumulated fatigue can be computed as the ratio of the total operating time to the associated creep fracture time (see Figure 4.6.3.1). When this ratio reaches the value of 100 %, intensified assembly component inspections are to be instituted. In general, operating temperature and operating loads change during operation. The total accumulated fatigue due to time-dependent creep can therefore be represented only as the sum of individual portions during constant time portions.

Accumulated Stress Cycle Fatigue (HP, IP Turbine Shaft)

For the shaft interior, the accumulated fatigue portions from the start-up processes and steady-state operation are totally cumulative over the entire operating time by means of the linear fatigue-accumulation hypothesis

(Miner-Palmgren law). If the permissible total accumulated fatigue is exceeded, intensified assembly component inspections are to be instituted. The material properties of the shafts, as well as any deficiencies of the forging that were within testing accuracy during the receiving inspection must be interpreted as incipient cracks, are included in the accumulated fatigue analysis by application of the methods of linear elastic fracture mechanics.

At cold starts the turbine shaft is particularly put at risk to reduced fracture toughness. In order to take care of this, an extra temperature simulator for the shaft centerline temperature is used to derive the permitted limits for the shaft body on the basis of fracture mechanics and to allow starting for the reduced fracture toughness of shaft materials at low temperatures. A transfer function continually modifies the specified stress limits for the shaft until the temperature has reached the specified level of full fracture toughness. This function can be dedicated to any shaft if technological reasons require to do so. The printout for the accumulated fatigue of shafts contains information of the mean integral shaft temperatures and various accumulated fatigue portions.

Turbine Stress Monitor

All turbine stress-free margins are displayed in bar-form on the color monitor of the turbine stress controller. The selected operating mode (fast/normal/slow) with digital turbine stress controller is indicated above the temperature free margins. The corresponding axial shaft temperatures are displayed above the shaft assembly components. The cyclic program processing is continuously indicated by a flashing cursor. In the event of a fault, a text message appears and the faulty assembly component is marked in red lettering.

Operation and Operational Data Evaluation

During speed operation the actual speed is displayed. The possible rate of speed change based on the minimum TSC free margin, is displayed below the actual speed. If the TSC function controls the reference limiter of the turbine controller or if the manual gradient setter is in control, a respective text is overlaid above the gradient display. If the TSC function is in control, the color of the free-margin bars of the controlling assembly component changes from green to yellow. If a negative free margin is calculated, the bar color changes to red. A signal indicates the actual power instead of the actual speed. Instead of the possible rate of speed change, the possible rate of power change for increasing and decreasing power is now displayed. As in speed operation, information is also provided here on whether the TSC influence is in control. The assembly component in control is obvious from the smallest displayed temperature free margin.

Extension Stage of an Automatic Start-Up System

Formation of the Start-up Criteria

In this stage the start-up criteria are formed and the saturated steam characteristics are computed. For the essential steps of turbine control (opening of stop valve, rolling off the turbine on steam, accelerating to nominal speed and applying load), the sliding temperature criteria are interrogated. These criteria determine whether the steam superheating is adequate or the steam temperatures are optimal for continued operation.

Determination of the Optimum Main Steam and Reheat Steam Temperature

The turbine stress controller determines thc main steam and reheat steam temperatures that are advantageous for smooth start-up of the turbine. From the representative turbine temperatures and the permissible temperature differences, the turbine stress controller determines the steam temperature reference values that are optimal for the turbine.

Monitoring of the Main Steam Line

With the stop valves opened, the main steam line is to be heated up to the point where the conditions for rolling off the turbine on steam are satisfied. Those conditions include adequate superheating and a sufficiently high main steam temperature relative to the metal temperature of the turbine. This heating up of the main steam line is to be carried out as subject to the heat-up free margins of the representative assembly components in the main steam line (e.g. main steam slide valve) of the turbine zone and of the monitored turbine assembly components. The smallest free margin is determined by a minimum value selector and is sent to the heat-up controller as the reference variable.

Software and Hardware

The computation of the accumulated fatigue of the assembly components as well as the associated long-term data storage operations is executed by a PC. The functions of the turbine stress controller are divided between different units of equipment corresponding to the requirements. The following functions are deposited in the functional module of the Automation System (AS):

▷ Wall-temperature monitoring,

▷ Formation of starting-criteria,

▷ Computation of the optimum main steam and reheat steam temperatures,

▷ Temperature monitoring of assembly components in the main steam line.

The turbine stress controller has an interface to the plant bus system for communication with other hardware and with the operating and monitoring system.

4.7 Seal Steam and Gland Sealing System

The shaft glands of the turbine must be sealed against the outside. The seal steam and gland sealing system prevents blowing of shaft seals, i. e. steam leaving or air entering into the turbine or condenser via the turbine shaft glands regardless of the load condition.

4.7.1 Seal Steam System

This system prevents steam leaving or air entering into the turbine or condenser via the turbine shaft glands regardless of the load condition.

Function of the System

The function of the seal steam system is depicted in figure 4.7.1.0. A seal steam header connected to all shaft seals is provided for this purpose. Pressure in the seal steam header and thereby in the shaft seals is controlled by either drawing off steam via a seal steam leak-off valve or by supplying additional steam via a seal-steam supply valve, as appropriate for the operating conditions at any given time. The excess steam is released into the condenser via the leak-off valve. Pressure is held at about 35 mbar (gauge) in this process. The stem leak-off steam tapping of the main steam and reheat steam stop and control valves is connected to the seal steam header. A warm up valve is provided to prevent the shaft seals for inadequate steam temperatures during the various load conditions.

Pressure Control of the Seal Steam

The seal steam pressure control system comprises an electronic controller acting on actuators which operate the leak-off and seal steam valves. The pressure in the seal steam header is acquired by pressure transducers. One transducer supplies the actual value for the control and the other the signal for display in the control room. The pressure transducers measure the differential between pressure in the seal steam header and atmospheric pressure. A pressure controller with PI action activates the actuators for the seal steam supply valve or leak-off valve on the basis of the result of the comparison between setpoint and actual value for pressure. To increase the availability of the seal steam pressure control provision is made for manual control from the control room. Changeover to this takes place manually from the control room, or automatically in the event of faults in the electrical part of the seal steam pressure control. The task of the electronic controller is to maintain the steam pressure in the seal steam header at constant pressure by appropriate actuation of the seal steam supply and leak-off valves. This control system ensures reliable sealing of the shaft seals under all operational conditions with minimum steam loss.

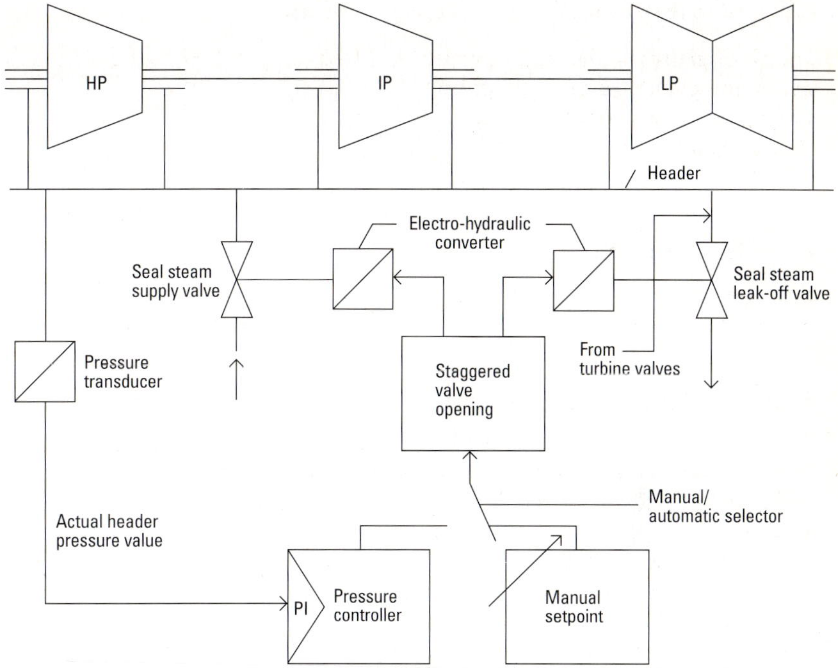

Figure 4.7.1.0 Functional Diagram of Seal Steam Pressure Control

4.7.2 Gland Sealing System

Function of the System

To prevent blowing of shaft seals of the turbine, a slight negative pressure in the leakage steam system is provided by an exhauster. The air fraction of the gland steam is removed from the header by an exhauster. The pressure in the gland steam chambers of the shaft seals can be adjusted by operating the valves at the suction fan or by setting the flap valves in the branch lines from the gland steam chambers. The steam fraction condenses in the gland steam condenser. The condensate is drained into the condenser via a drain trap.

4.8 Bypass Station

For reaching a high operational flexibility of a power plant, to optimize start-up and shut-down of the plant and to control normal and abnormal conditions, it is useful to operate the boiler independently of the steam turbine [6]. External HP and LP bypass stations allow independent boiler and turbine operation during start-up, shut-down and any transient conditions, such as sudden load rejections. When starting up a unit, the bypass systems enable the boiler to be operated prior to the turbine in order for the boiler steam outlet temperatures to be matched with the turbine component metal temperatures before steam is admitted to the turbine. Cycling operation results in high metal temperatures in the turbine during overnight shut-downs. This is an advantage for a fast restart to full unit load in the morning when matching the turbine metal temperatures with the main and reheater steam outlet temperatures. Completely independent boiler and turbine operation can also be achieved for full or partial-load rejection depending on the size of the bypass station. This decoupling of boiler and turbine is realized by the bypass system. External HP and LP bypass stations operate in conjunction with the turbine control. Figure 4.8.0 shows an overview of the HP/IP bypass system. The bypass system consists of:

- ▷ HP bypass station; Bypass from main steam line to cold reheat line and reheater (bypass of the HP turbine section),
- ▷ LP bypass station; Bypass from hot reheat to the condenser (bypass of the IP and LP turbine section).

Both bypass stations are equipped with water injections to control the steam temperature and with fast-open functions for unit malfunctions (i.e. fast opening of the bypass station in case of a turbine trip).

4.8.1 HP Bypass Station

The HP bypass station leads the steam from the main steam line to the cold reheat line bypassing the HP turbine in case of hot or cold start-up of the boiler and on load rejection (turbine trip) and in the event of overpressure in the main steam line. The bypass station consists of the HP steam-converting valve, the spray-water control valve and the spray-water isolating valve. The steam-converting valve carries out simultaneously steam pressure and temperature reduction by means of injection of spray-water. The single seat, which has a throttle cone with a perforated cage trim, is tight-closing. The HP bypass steam is desuperheated by means of feedwater injection via a spray-water control valve. The spray-water control valve with a multi-stage valve seat is designed for a high pressure drop and offers a high degree of resistance to wear and erosion.

The task of the HP bypass control is to control the main steam pressure in the event of malfunctions. The pressure setpoint can be adjusted to realize the pressure conditions. During normal load operation the controller adjusts the pressure setpoint to a certain margin above the actual pressure which corresponds to the momentary turbine load. This prevents an unnecessary opening of the HP bypass during pressure changes that are caused by load changes. If the actual pressure rises more than the adjusted pressure setpoint, the HP bypass valve opens. In case of high pressure deviation or turbine trip, the HP bypass station is equipped with fast-open functions, which open the HP bypass valve completely and quickly.

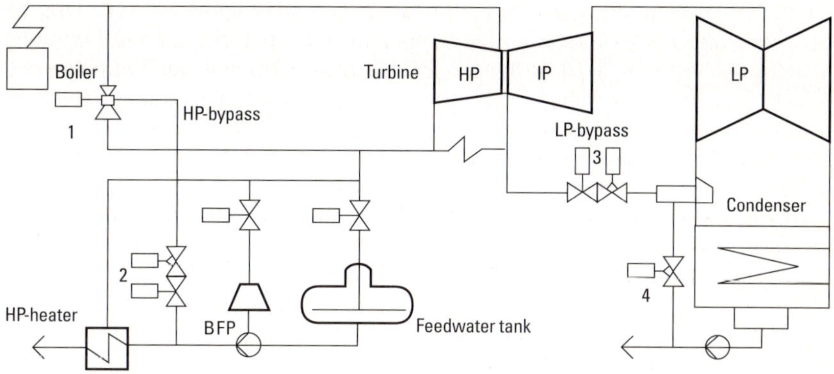

Figure 4.8.0 Overview of HP/LP Bypass System

The task of the LP bypass control is similar to the task of the HP bypass control. The LP bypass control realizes the desired pressure conditions for start-up. It assures sufficient steam flow through the reheat system to cool the reheater by a minimum opening condition. The HP bypass valves are motor driven.

4.8.2 LP Bypass Station

A signal from the bypass controller activates the electrohydraulic bypass stop and control valve to feed steam from the reheater system into the condenser bypassing the turbine. The bypass stop valve acts as an isolation valve being either closed or fully open. The bypass control valve, on the other hand, regulates the flow of steam to the condenser according to the signals from the bypass controller. One stop and one control valve are combined in a single casing. The discs and stems of the stop and control valves are manufactured from one piece. A connection to the condenser is provided for the combined bypass stop and control valve. Due to an orifice provided in this line, a second critical expansion takes place downstream of the bypass valve. This rules out inadmissible stresses on the condenser during subsequent supercritical expansion.

The LP bypass steam is desuperheated by means of condensate injection via a spray valve in order to protect the condenser against inadmissible temperatures. A control valve controls the flow such that the injected water flow matches the steam flow even on fluctuating condensate pressure. A part flow can be set by means of a manually-operated valve in the event that this control valve fails. The stop valve section is fitted with a basket-type strainer to catch any contamination. The supply of control fluid for the hydraulic actuator of the LP bypass valve is by means of the hydraulic supply unit. Control valve movement is initiated by the position controller in the event that the actual valve position deviates from the setpoint by sending a signal to the electrohydraulic converter. For rapid closing operation resulting from load rejection, two two-way valves actuated by solenoids are provided. Due to the design features of the two-way valves, short closing and opening times of less than 150 ms can be achieved. The two-way valves are standard cartridge valves with piston guidance.

4.8.3 LP Bypass Controller

The bypass controller controls the pressure in the reheat system. Under certain operating conditions during start-up, shut-down, and load rejections excessive steam not being expanded in the IP and LP turbine is bypassed to the condenser by opening the LP bypass valves. The bypass control system consists of an electronic controller controlling the LP bypass valves and another electronic controller controlling the water injection valve. The link between the electronic controllers and the hydraulic equipment are electrohydraulic converters. The structure of the LP bypass controller is depicted in Figure 4.8.3.0.

The bypass controller has to maintain the pressure in the reheat system constant corresponding to the preset reference value. The setpoint is derived

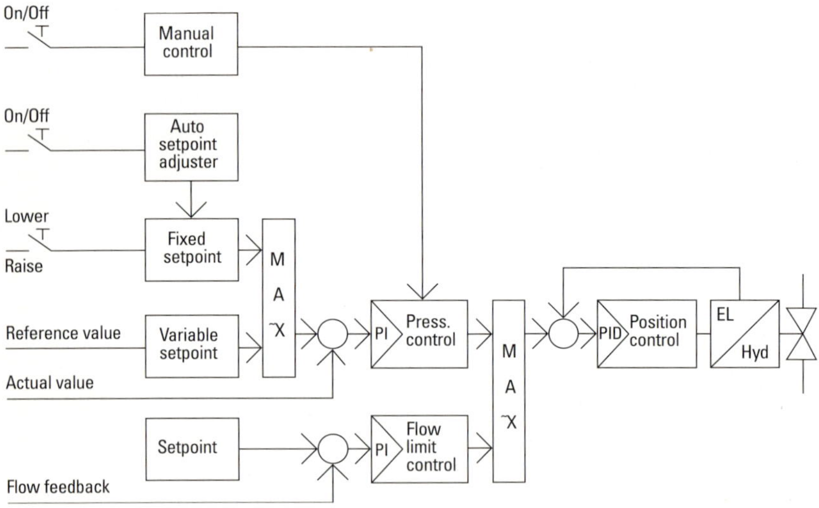

Figure 4.8.3.0 LP Bypass Control Structure

from the pressure at the first stage blading of the turbine which is proportional to the steam flow. During start-up and shut-down, this variable setpoint is replaced by a fixed reference value. The actual value is the pressure measured in the hot reheat line. The control deviation between reference and actual value is fed to the PI-pressure controller. The pressure controller output compared with the actual bypass valve position is fed to the PID-position controller. The position controller output is directed to the electrohydraulic converter which in turn actuates the control valves.

In case of bypass operation first the water injection valve is opened to supply injection water to the condenser inlet. Then the bypass stop and control valve is opened to bypass excessive steam to the condenser. If, however, injection water is not available or the vacuum is low, the bypass operation is interrupted.

The bypass controller can be switched over from automatic to manual control either by the operator or automatically upon failure of the automatic controller. In this case the bypass valves can be controlled by the manual reference setter. To achieve a bumpless transfer from automatic to manual control, a matching controller is provided. The LP bypass controller can provide an anticipatory control signal for operator's use to pre-control the reheat safety valves, if so required.

4.9 Turbine Drain System

If the wall temperature of the components is lower than the saturation temperature of the steam flowing through them, condensate is produced in the turbine casings, piping and valves. Condensate removal is provided by drain valves to prevent interference with operation or even damage to the turbine plant due to insufficient draining. Such adverse effects of inadequate draining may be, e.g.:

- ▷ Distortion of casings due to one-sided cooling by accumulating condensate,
- ▷ Damage to piping fixtures and changes in turbine alignment due to cavitation in pipes,
- ▷ Irregularities in seal steam control due to condensate accumulation in the header.

The drain valves are opened and closed as a function of the turbine component wall temperatures or the status of the swing check valves in the extraction lines. The drain valves in the region of the main and reheat steam valves are also used for warming up the main and reheat steam valves. Drain valves having the same control criteria are actuated simultaneously in groups. The condensate passes through drain valves via headers into the condenser. The drain valves are mounted on the headers depending on the pressure (the highest pressure is nearest the end of the header). The flash-box into which the headers discharge are equipped with spray attemperators to limit the temperature. The drain valves are provided with, e.g. pneumatic actuators.

4.10 Oil Supply System

The oil supply system, consisting of main oil tank, several oil pumps, valves and interconnecting piping, supplies the required volume of oil at the necessary pressure to the points of consumption:

- ▷ Pressure for bearing lubrication and cooling (5.5 bar),
- ▷ Pressure for bearing lubrication on emergency rundown (2.5 bar),
- ▷ Pressure for jacking oil (160 bar),
- ▷ Pressure for shaft turning gear (160 bar).

The main oil tank serves not only as a storage tank but also for deaeration of the oil. The capacity of the tank is such that the full oil inventory is circulated no more than 10 times per hour. This results in a retention time of about 6 minutes from entry into the tank to suction by the pumps. This time allows for sedimentation and air detrainment of the oil. Reserve capacity is provided between the normal operating level (corresponding to rated capacity) and the tank cover to accept oil from the entire oil supply system when the turbine generator unit is shut-down. The tank is constructed with a sloping bottom with drain connections at the lowest point. Oil returning to the tank from the oil supply system first flows through a submerged inlet into the riser section of the tank, where the first stage of deaeration takes place as the oil rises to the top of the tank. Oil overflows from the riser section through an oil strainer into the adjacent section of the tank and then flows lengthwise through the tank until it is drawn off at the opposite side by the oil pumps. The main oil tank has a local oil level indicator and level switches which signal when maximum and minimum levels have been reached. Figure 4.10.0 shows the simplified layout of the lubrication oil system.

The following submerged pumps are mounted on the tank:

- ▷ Two 100% single-stage main oil pump (centrifugal pump),
- ▷ Single-stage emergency oil pump with DC motor (centrifugal pump),
- ▷ Jacking oil pump with three-phase AC motor (screw pump).

Modern steam turbine generator sets are equipped with a system providing electrically driven main oil pumps, which has advantages over a direct shaft driven oil supply pump.

4.10.1 System Configuration

Two three-phase AC main oil pumps and one DC driven emergency oil pump are provided. The lube oil delivered by the main oil pump is fed to the bearings via the coolers, temperature control valve, duplex filters with change-

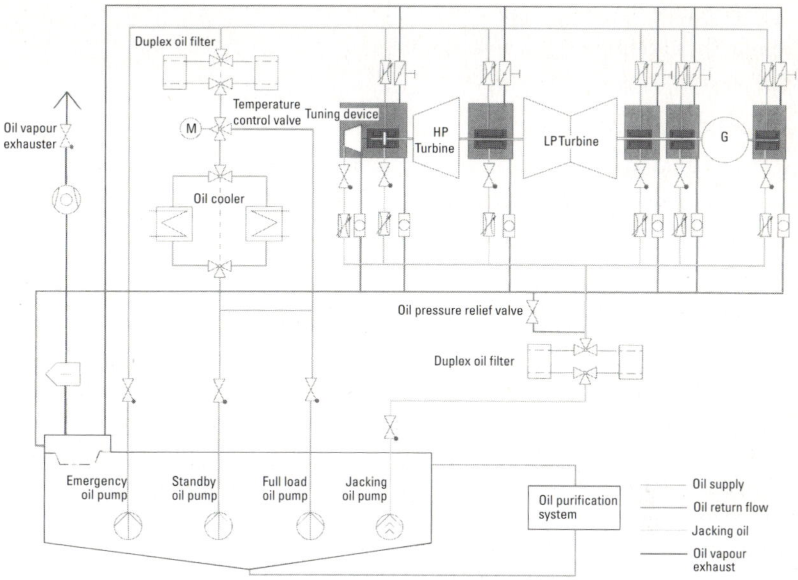

Figure 4.10.0 Simplified Layout of the Lubrication Oil System

over capability, and bearing oil flow restrictors. In emergency pump operation, the oil is fed ahead of the bearing oil flow restrictor, bypassing the coolers, the temperature control valve and the filters. For this reason, the emergency oil pump is designed for a lower service pressure. One of the two electrically driven main oil pumps is on-line at any time, while the second main oil pump remains on standby. If the on-line pump develops a fault, the standby oil pump is cut in as the pressure in the lube oil system drops below a pre-set pressure threshold.

The criteria for cutting in the standby pump and the emergency oil pump ensure that the required oil pressure is maintained during all changeovers between pumps and that thus the oil supply to the bearings is under control at all times. The lube oil supply is not interrupted during any phase of operation. The two main oil pumps are connected to separate power supplies via physically separated cable runs. In the event of turbine trip, the jacking oil pump is also cut in as a function of turbine speed. In the event of a failure of the on-line oil pump, it may under certain circumstances be possible to dismantle and repair the failed pump on site or replace it with a pump from the spare parts store. The ability to separate the oil supply from the shaft-driven main oil pump offers several advantages:

- ▷ Enhanced operating reliability and availability of the turbine plant,
- ▷ Lower power losses,
- ▷ Less installation effort and easier servicing,
- ▷ Shorter overall length of the turbine,
- ▷ Better running behavior and less noise.

4.10.2 Oil Supply during Turbine Generator Set Operation

During turbine generator set operation one of the two electrically driven full-load oil pump supplies the required volume of oil to the bearings. The necessary oil inlet temperature for the bearings of the turbine generator set is ensured by the thermostatic oil mixing valve installed in the oil line downstream of the oil cooler. This three-way valve controls the oil flow which bypasses the oil cooler.

A jacking oil system is provided to lift the line of shafting during turning gear operation and to supply the hydraulic turning gear with motive oil. A low volume flow of high pressure oil is forced under each of the rotor journals to lift the rotor off the bearings in order to prevent metal-to-metal contact at low speeds. By means of an oil filtration loop, the oil is continuously passed through switchable duplex micro filters and supplied to the bearings. The jacking oil pump located in the main oil tank is a positive displacement pump driven by a three-phase AC motor. The required system pressure is kept constant by means of a pilot operated pressure limiting valve. The required jacking oil pressures are set for each bearing by means of precision control valves upstream of the individual bearings.

All spaces above the oil level in which oil vapor can form (oil tank, return line, bearing pedestals) are exhausted by a common fan. The fan produces a pressure slightly below atmospheric pressure. An oil separator is installed on the suction side of the fan and the air is being discharged into the atmosphere.

The shaft-turning gear is connected to the shaft jacking system; this means that the turning gear starts moving as the shaft jacking system is put into operation. An adjustable throttle valve in the feed pipe of the hydraulic motor is used to set the turning gear speed. The turning gear drives the turbo generator set continuously at about 60 rpm. The advantage of this relatively high turning speed is the more uniform temperature distribution in the turbine components during start-up and cooling-down procedure. Since the turning gear device operates with the oil pressure of the lift oil system, it is ensured, that an effective oil pressure is established before turning gear operation starts.

In general, two 100 % capacity oil cooler are provided. The oil coolers are of shell and tube type. The nozzles for the cooling water inlet and outlet are at the bottom of the cooler. The water box at the top reverses the flow of the cooling water. Separate flanges with appropriate gaskets are provided for maximum possible security against leakage of water into the oil side. Heat is exchanged via plain tubes. Various internals within the oil cooler force the oil to flow in a cross-flow pattern, thereby ensuring uniform and high heat transfer.

4.11 Generator Coolant Temperature Controller for Generators with Water-Cooled Stator Windings

To keep stressing of the generator materials to a minimum, changes in the temperature of the generator components should be kept as small as possible even during load changes [7]. This is best done by maintaining the coolant inlet temperature constant.

To fulfill this task, the hydrogen (H_2) temperature controller controls the flow of cooling water through the H_2 cooler. In generators with water-cooled stator windings, the primary water temperature controller controls the flow of cooling water through the primary water-cooler.

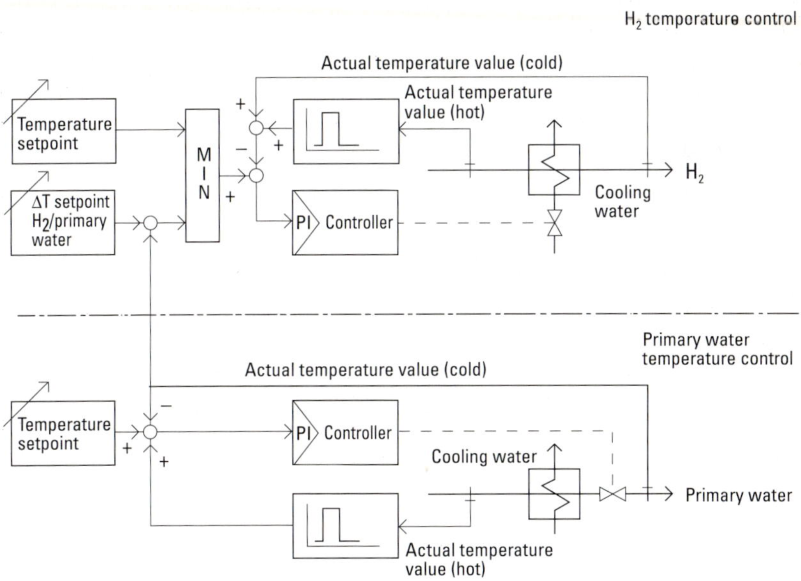

Figure 4.11.0 Functional Diagram of Generator Coolant Temperature Control

Both controllers act as PI-action controllers. In addition to the adjustable setpoint and the cold gas or cold primary water (inlet) temperature, an adjustable hot gas or hot primary water (outlet) temperature component is input to the controllers as feed-forward to accelerate their response to load changes. The functional diagram of the control systems is shown in Figure 4.11.0.

4.12 Excitation System

4.12.1 Brushless Excitation

For large steam turbine generator sets the brushless excitation system is in use and has proven effective for many years of operation [2]. The brushless excitation system has the advantage of easy transmission of high generator field currents without carbon brushes and slip rings and guarantees therefore a high operating reliability with practically no maintenance. A full ceiling voltage is provided at all times when auxiliary power is available. The stationary poles of a three-phase main exciter are excited by a pilot exciter via a voltage regulator. The three-phase current generated in the rotating armature of the main exciter is rectified via rotating diodes and conducted to the rotor winding of the turbogenerator via copper conductors in the shaft center. Figure 4.12.1.0 shows the block diagram of the system. Within the mechanical coupling between the exciter set and generator, a specially developed multicontact system establishes the connection to the DC supply leads. The THYRISIEM®D voltage regulator is a two-channel design and consists mainly of components of the digital control system SIMADYN D.

4.12.2 Static Excitation

In addition to brushless excitation, static thyristor converter excitation is also in use particularly for smaller units. The THYPRIPOL®D voltage regulator for static excitation systems is also digital and is a two-channel design [2]. Together with one or two power sections, the digital voltage regulators THYRISIEM D and THYRIPOL D form a modular product family.

Siemens/KWU gas turbine generators are primarily started up with the aid of static start-up frequency converter. For run-up with a static frequency converter, the generator rotor has to be excited with the aid of slip rings also from standstill. For this reason, gas turbine generators are normally provided with a static excitation system. The excitation system and start-up frequency converter are supplied as a compact unit featuring digital technology (see Figure 4.12.2.0).

4.12.3 Generator Voltage Regulator

Generator voltage is controlled by changing the excitation current (functional diagram see Figure 4.12.3.0). Two control systems are available for this

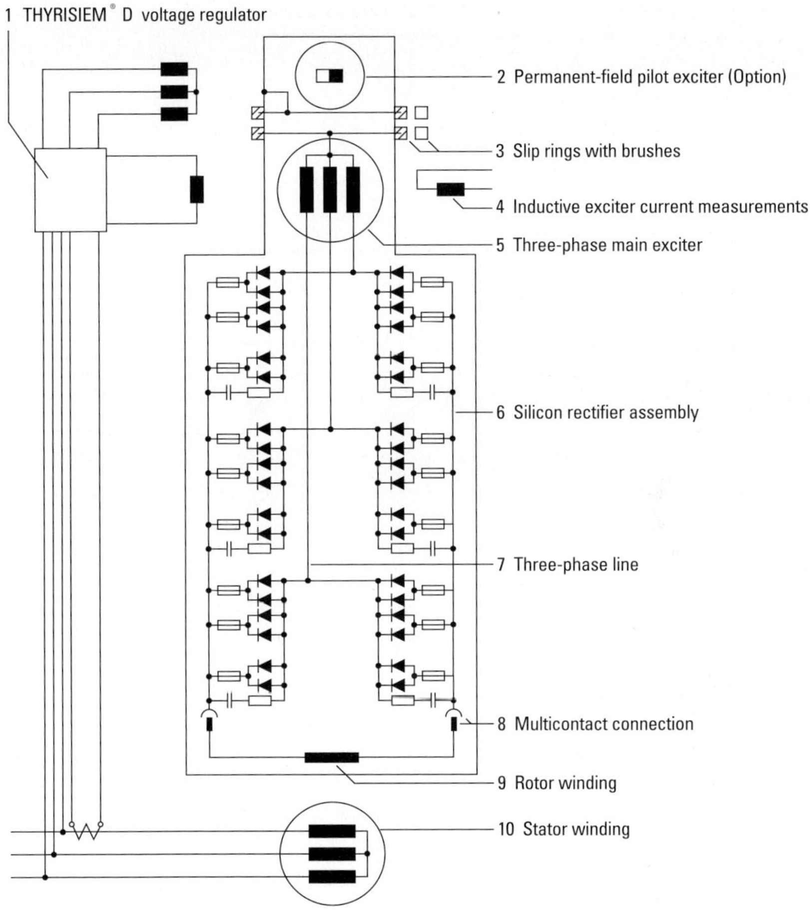

Figure 4.12.1.0 Brushless Excitation, Blockdiagram

purpose [2, 8]. Control system 1 (automatic control) has the following functions:

▷ Control of generator voltage to match the setpoint by changing the excitation current,

▷ Field-forcing limitation; instantaneous limitation of excitation current to the maximum allowable current usually 1.4 to 1.5 times the rated excitation current. Protection of equipment and machinery against overloading due to excessive overexcitation,

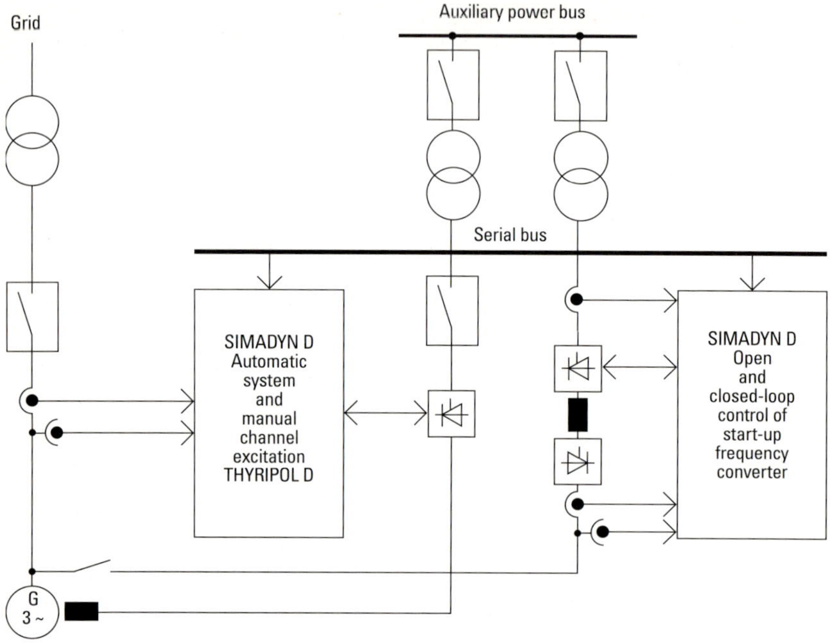

Figure 4.12.2.0 Static Rectifier Excitation and Start-Up Frequency Converter

- ▷ Overexcitation limitation; delayed reduction of excitation current to rated current in the event of overexcitation. Protection of equipment and machinery against overloading due to prolonged overexcitation,
- ▷ Underexcitation limitation; assurance of minimum excitation for stable parallel operation of the generator with the grid or limitation of reactive power below underexcitation conditions. The response characteristic is derived from the generator variables, reactive power, active power and terminal voltage.

Control system 2 (manual control) controls the excitation current on the basis of the manually input reference. This is used to record generator characteristics during commissioning and major inspections and to operate the generator in short circuit for the purpose of establishing settings for protective actions. It is also used to select generator excitation, should the automatic voltage controller fail. During normal operation and during start-up and shut-down of the turbine generator unit, generator voltage is controlled by control system 1 (automatic control). The reference of control system 2 (manual control) is matched automatically so that changeover to the excitation current control mode can be effected either manually or automatically at any time.

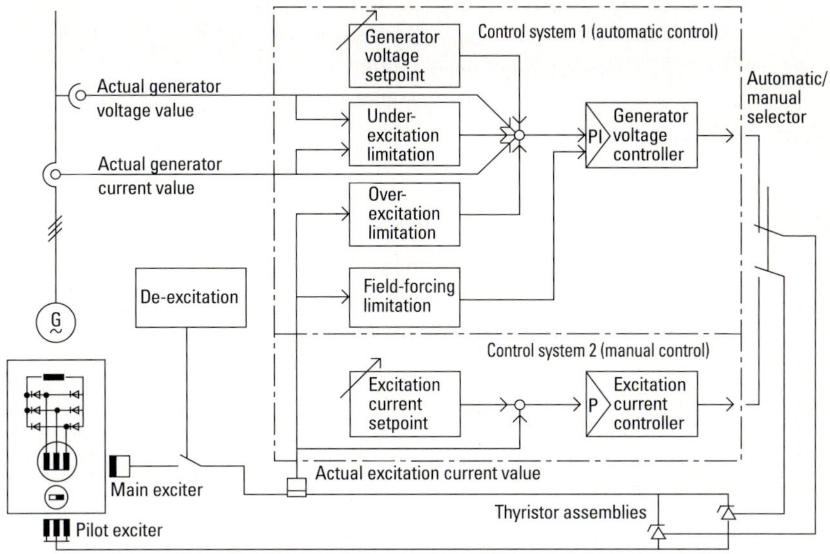

Figure 4.12.3.0 Functional Diagram of Generator Voltage Control

4.13 Open-Loop Control at the Turbine Generator Unit

Depending on the degree of automation, the technological processes are controlled by automatic functional group controls. Reasons for automatic functional group controls are [4]:

▷ To save time and costs and achieve low-stress start-ups and operation of the turbine generator,

▷ To avoid maloperation, i. e. to keep loss of availability as low as possible,

▷ To relieve the work load on the control room personnel.

The task of the turbine control functional group is:

▷ Automatic, low-stress start-up of the turbine according to the turbine stress evaluator (cold start or warm start) from turning operation to rated speed,

▷ Automatic synchronization with the grid,

▷ Loading the turbine generator with minimum load,

▷ Low-stress shut-down of the turbine,

▷ Subloop control, e.g. for the turbine oil supply or for automatic draining of the turbine during start-up,

▷ Program stop during impermissible operating conditions.

Independent subloop controls are provided to realize the start and loading procedure of the steam turbine generator set. Figure 4.13.0 gives a general view of the function diagram as an example.

4.13.1 Functional Groups

Turbine Generator Set

The task of the functional group turbine is:

- ▷ To start-up and shut-down the turbine generator unit automatically,
- ▷ To load the turbine generator up to the set power reference,
- ▷ To perform switching operations required for operational reasons or for fault elimination.

Functional Groups of the Turbine Auxiliary Systems (Examples)

Oil Supply

This functional group controls the systems concerned assuring the turbine generator unit oil supply. Independent subloop controls are provided for the main oil pumps and for the emergency oil pump. The pumps are cut in and out as a function of the oil pressure. The subloop control stops turning-gear operation when the turbine generator unit has clearly exceeded turning-gear

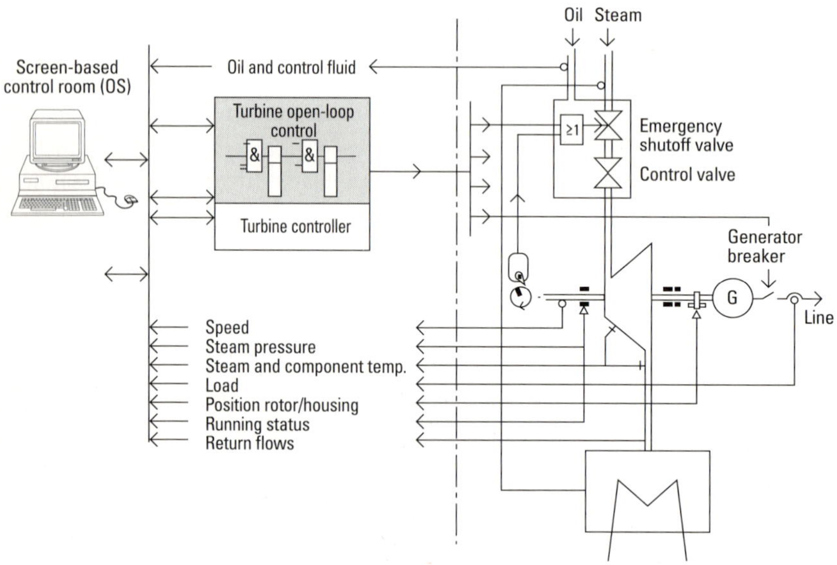

Figure 4.13.0 Overall Structure of Turbine Open-Loop Control, Example

speed and restores it before turning-gear speed is reached as turbine speed drops. The shaft jacking subloop control keeps one jacking oil pump continuously in operation during turning-gear operation, cuts the pump out when the hydraulic turning gear is disengaged, and restarts it together with the hydraulic turning gear as turbine speed drops.

Drain System

The drain valves are opened and closed as a function of the turbine component wall temperatures or the status of the swing check valves in the extraction lines. Drain valves having the same control criteria are actuated simultaneously in groups.

Condensation

The functional group condensation comprises all equipment which is used directly to evacuate the air from the turbine condenser and to seal the turbine shafts. Before the seal steam controller is cut in, there are a number of conditions that must be fulfilled. Vacuum pumps with air ejectors evacuate the air from the condenser. First the vacuum pumps and the service water supply are switched on simultaneously. The air exhaust valve remains closed until the vacuum pumps have built up a certain vacuum. When the air exhaust valve is opened, air evacuation from the condenser is performed by the vacuum pumps only. As soon as the steam flow into the condenser has ceased during turbine shut-down, all vacuum pumps are cut out and the solenoid valve to the ejectors and the air exhaust valve are closed.

4.13.2 Function of the Open-Loop Control

The turbine open-loop control consists of the following:

▷ Analog signal conditioning,

▷ Criteria generation,

▷ Signal gating,

▷ Command output,

▷ Criteria and step display.

The switching operations required for each of the above mentioned processes are executed automatically in a sequence control after the necessary criteria have been logic-gated. The commands for the actuators are generated by the command output; the control room personnel receives the relevant information via the criteria and step display. Criteria-dependent step enabling conditions (e.g. warming of the turbine, vibrational behavior of the rotor, etc.) ensure optimum adaptation to each plant condition; rigid start-up curves against time are thus avoided resulting in the shortest possible start-up times. Figure 4.13.2.0 shows the function sequence.

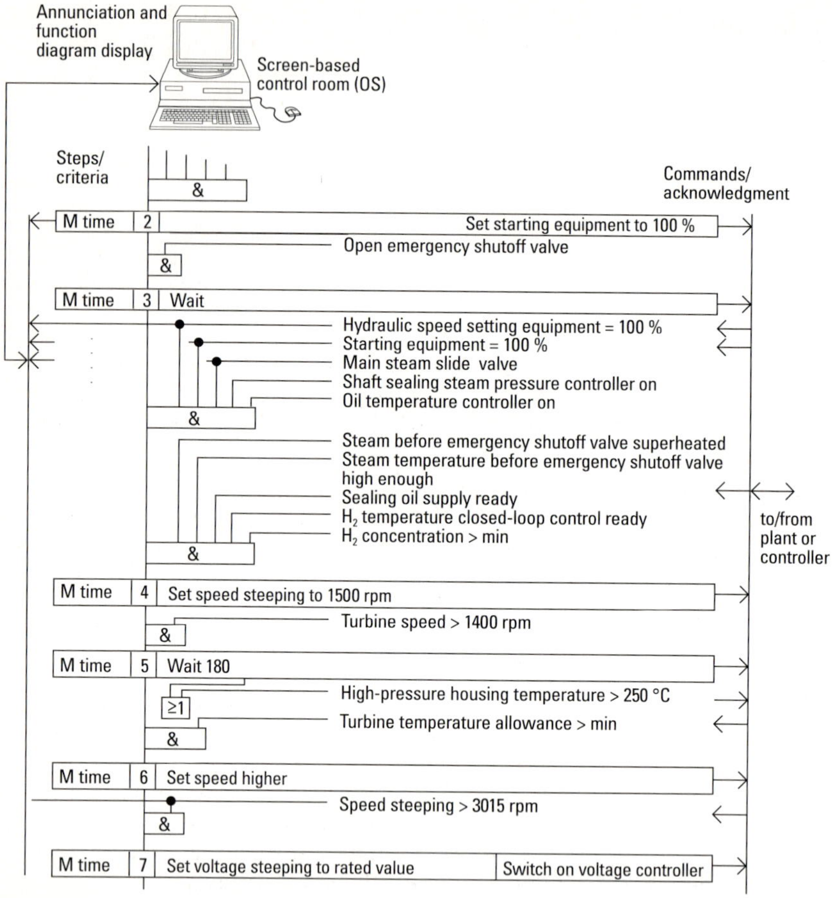

Figure 4.13.2.0 Starting Program Open-Loop Control, Example

5 Diagnostics System DIGEST

5.1 Introduction

Thermodynamic optimization of operation, recognition of incipient faults during power plant operation are important factors in achieving the greatest degree of availability and cost-effectiveness in a power plant. The step-by-step introduction of condition-oriented maintenance of power plant components, could turn out to be a promising method for reducing maintenance costs.

Today, a high degree of automation and extensive I&C equipment are required to achieve these goals. However, this concept involves a high concentration of information, placing great demands on the operating personnel in their process monitoring and control tasks, especially in the case of assessing transient events and incipient faults.

Use of diagnostics systems in power plants yields considerably enhanced information consolidation. The implementation of high-performance, front-end computers, whose computing capacity approaches that of the main frame computers in the past and modern programming methods are at the heart of this concept. Large-scale acquisition of on-line operating data, signal exchange via bus systems, and broad-scale data comparison now make it possible to perform a detailed diagnosis of plant status in the form of verbal statements and sophisticated graphic displays. The resulting comprehensive overview of the momentary plant status facilitates optimum and thus cost-effective operation. It also makes it possible to recognize incipient malfunctions in good time and to take remedial action. Diagnostics as an information system to relieve the load on the plant operating crew is incorporated at the operator control level of the I&C hierarchy. This concept can also be applied, subject to digitalization of the signals, in plants with "hard-wired programmed" I&C.

5.2 Basic Concept

Monitoring leaves the essential task of assessing the status readings to the operating personnel. However, the use of workstations and newly developed PC's in power plants now makes it possible to acquire and handle a large volume of data and thus achieve great information density. Monitoring need no longer be limited to the display of discrete readings or the identification of unusual conditions. DIGEST (Diagnostics Expert System Turbo-machinery)

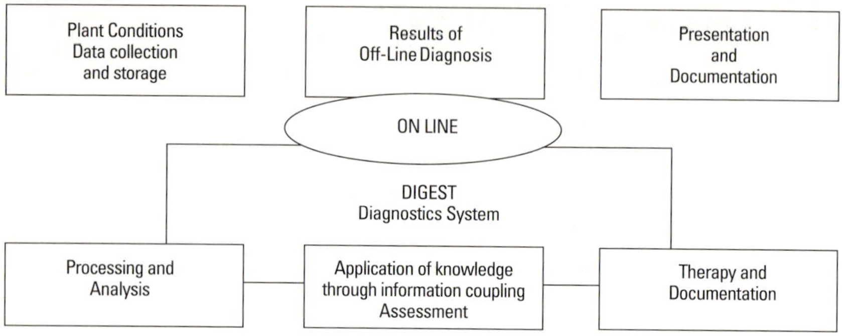

Figure 5.2.0 Diagnostics System DIGEST, Components and Processes

annunciates threshold violations and permits information to be derived on-line as to the current and future status of the plant. The basic function of an on-line diagnostics system is thus to make it possible to gather comprehensive parameter readings and other status information during plant operation and to combine these readings and information into an overall assessment of the operating status of the plant. Expert Systems carry out machinery diagnoses – often by extensive analyses – to recognize incipient irregularities in machinery and processes. Figure 5.2.0 shows the structure of the system. The various fields of activities of DIGEST results in diagnostics to achieve the following goals:

▷ Service reliability and high availability (recognition of incipient faults and condition-oriented maintenance),

▷ Optimal thermal operation,

▷ Monitoring of the component service life.

We distinguish between on-line and off-line diagnostics as follows:

- On-line diagnostics, computer-assisted assessment of parameter readings (during operation) for the purposes of early recognition of incipient faults, identification of their causes, finding suitable countermeasures.
- Off-line diagnostics, it enables processed, stored and measured data to be analyzed at any time. May also be performed during turbine shut-down, e.g. by inspection of certain components without opening up the turbine and by measurements and tests performed on the turbine during standstill.

An accurate diagnosis of faults requires a thorough understanding of the design and operation of the unit. It must also be complemented by the operating experience with the unit concerned. For certain tasks the results of the diagnostics system are not required immediately, but can be provided at a later date. It is also possible to obtain supplementary information from the off-line diagnoses. This can be done interactively with the diagnostics system in dialog mode. The purpose of an on-line diagnostics system is to acquire and process large amounts of data on a unit and to present this as useful information concerning the operation of the machine to the plant personnel. The goal of such a system is to detect possible failures and to recommend certain specific actions to be taken. These can be summarized as:

- ▷ Recording of operating data and analysis of any changes,
- ▷ Linking various readings and data into operator messages,
- ▷ Comparing actual and specified data and determining trends,
- ▷ Identifying faults and their causes,
- ▷ Predicting any trends in the future operation of the plant,
- ▷ Providing recommendations to the operator on actions and maintenance activities.

The initiation of a diagnosis is event-oriented, i. e. it takes place automatically, far in advance of alarm annunciation, when the indicated values violate their permissible limits. In addition, a diagnosis can be initiated at a specific time or for manual evaluation of a problem area. Conditioned data are filed in either the long or short-term memory of the diagnostics computer, depending on the type of data. The data are available in the form of logs or graphs and can also be output on-screen and via the printer. All information, warning messages and recommendations derived from the diagnostics system is stored. The stored data pertaining to a specific fault can also be called up from disks or other data carriers. Teletransmission of fault data is also possible by modem. This makes it possible for the manufacturer to perform a detailed on- line “fault analysis”. Monitoring facilities identify the operating status of a turbine installation and/or annunciate deviations from the desired operating status.

If data transfer to the computer is not possible, the data can be entered manually. A concept in which the data are read in using a hand held computer, for example during a tour of inspection (on site), would permit direct data transfer onto the computer on which the system is run. This concept would also permit transmission of the measured values by modem to a central point in the power plant or for analyses by specialists at Siemens/KWU.

5.3 Structure of DIGEST

Definitions of the terms monitoring, analysis and diagnosis are summarized in Table 5.3.0. With regard to the modular design of hardware and software, and to the diversity of its areas of application, DIGEST is structurally divided into three process levels: One for data acquisition, conditioning and monitoring, one for module-specific analysis and diagnosis and one for centralized diagnostics. Figure 5.3.1 shows the structure of the system. The functions of each process level are shown by way of an example of a condenser fault evaluation in Figure. 5.3.2.

Table 5.3.0 System Task and Requirements

Tasks	Description	Requirements
Monitoring	Acquisition and presentation of operating conditions and comparison with specific data, alarms and limit data to prevent impermissible/ uneconomical operating conditions (VGB Regulation R103M)	Personnel with specialized technical knowledge
Data presentation	Visualization of data by trend display	
Analysis	Calculation of reference data relative to the operating condition as the basis for comparison between actual and specified data for condition evaluation	
Diagnosis	Automatic evaluation of data and analysis results by compilation and association for fault recognition	Personnel with specialized technical knowledge not required

Monitoring the usual measured data (DIGEST level 1) for a condenser does not show whether deviations exist that will lead to faults because the condenser pressure is dependent both on the cooling water condition and power. To detect faults, an analysis (DIGEST level 2), calculation of the target state has to be performed. The cause of faults can only be detected by calculating the target state, i.e. from the results of the analysis and applying expert knowledge. The knowledge interpretation is performed with a matrix, Figure 5.3.3, the "rules" derived from the matrix are stored in the DIGEST expert system [8].

During on-line operation, the central diagnostics expert system automatically reports any irregularity. The cause of the faults could only be recognized by

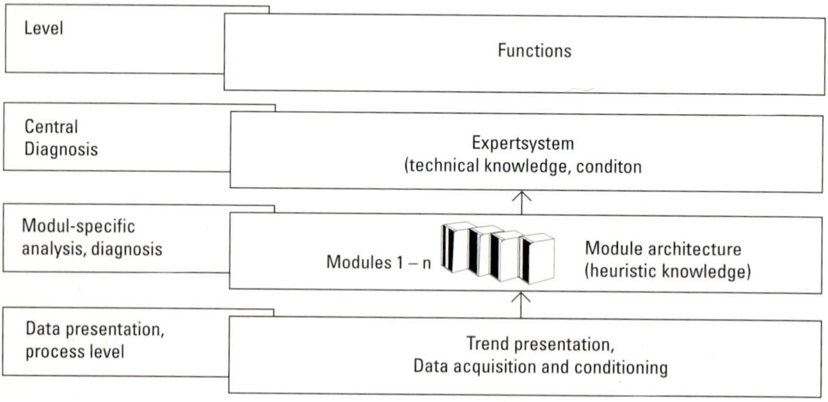

Figure 5.3.1 Hierarchical Configuration of DIGEST

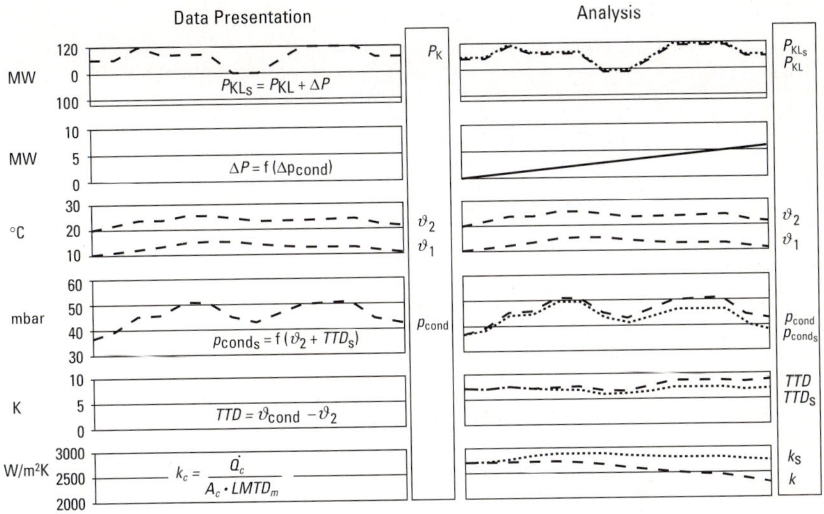

Index S = Target value, without index actual value other terms and indices see chapter 7

Figure 5.3.2 Condenser, Data Presentation and Analysis

associating the results of the analyses and applying special technical knowledge. Expert Systems carry out the individual diagnoses to recognize incipient irregularities in the process. The system draws logical conclusions. The results are not only one diagnosis but a sorted listing of all candidate diagnoses which attain a certain confidence level. The irregularities are evaluated by analyzing data and trends, and the signals are made available for "fuzzy logic" processing in the language of the expert. The expert system contains a very large knowledge base derived from failure statistics. Diagnostics systems have already become indispensable tools of power plant technology. Applications of the DIGEST system for on-line diagnosis have a useful role in this field.

5.4 Modular Architecture

DIGEST for steam turbine generators is of modular architecture. The various modules and module groups are based on the proven steam turbine plant functional groups. Table 5.4.0 shows an overview of the diagnostics system (overall concept) and its modular structure for a steam turbine generator. The diagnostics system comprises modules covering the diagnostics tasks for the turbine generator set as a whole (on the left-hand side in the table) and modules dedicated to specific components (on the right-hand side in the table). Extension of the scope of the modules provided for a given diagnostics task on the turbine generator or any of its components permits more in-depth

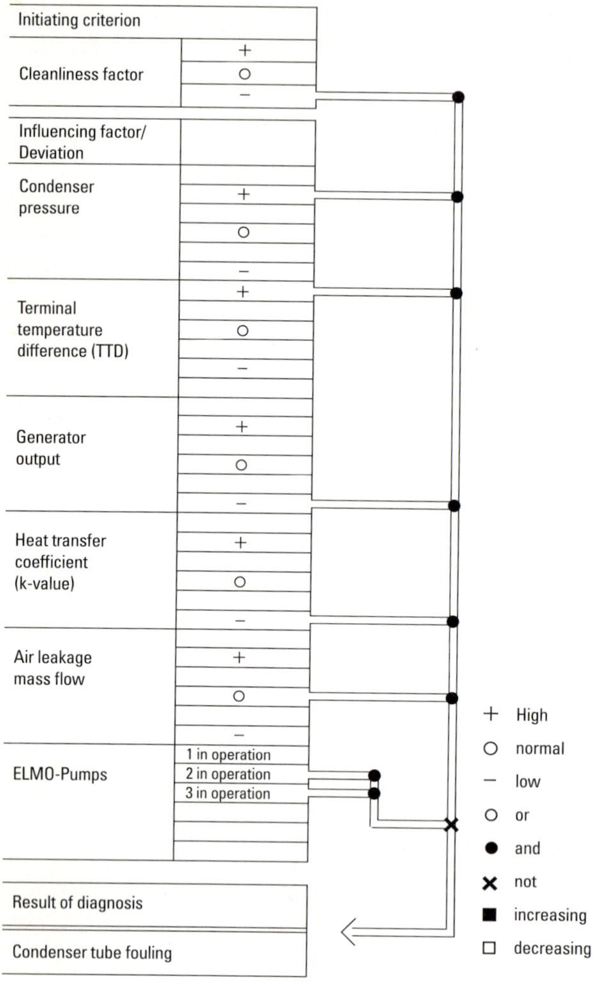

Figure 5.3.3 Condenser, Fault Diagnosis

diagnostics within that field. The modules are combined at a higher level so that their interaction and interrelations can be observed. Consistent and easy-to-use communication between the modules and the user of the system is ensured by a common user interface.

The selection of modules for further diagnostics enables that specific tasks extend the scope of diagnostics performed. The use of modular building blocks enables the diagnostics system to be developed with a high degree of

Table 5.4.0 Steam Turbine Generator Overall Diagnostics Concept DIGEST

Turbine Generator Set				Components and Auxiliaries		
Vibrations	Bearings	Life Expectancy	Thermo-dynamics	Turbine	Condenser	Generator
Frequency analysis, data base early crack detection	Journal bearing, thrust bearing, bearing oil	Low-cycle fatigue	Plant efficiency	Control valve actuators, valve operation	Thermodynamic monitoring	Cooling system
Normal data and limit comparison	Oil supply, pumps, oil cooling oil purification	Creep damage	Turbine generator set and feedwater heating system	Casing distortion, reverse flow, drains and extractions	Air extraction system	Conductor and insulation
					Condensate control	Hydrogen sealing
Failure analysis, trend monitoring, logic gating	Shaft turning gear, shaft Lift oil system	Long-term storage of operational data	Feedwater heaters, extractions, condenser	Exhaust steam temperature, valve lift	Tube cleaning	Moisture (dew point)
					Single tube	Cooler
Turbine plate deformation		Torsional loadings	Water/steam cycle components	blading	Circulation water cooling tower	Exciter system

flexibility for use in all types of power plants. Retrofitting such systems in operating units is also possible. Expansion of the diagnostics system can be accomplished in steps. If the base module comprising the fundamental program and the associated hardware are adopted first, additional modules can always be added on.

5.4.1 System Framework

When a module is first installed, a general system framework is supplied into which further modules can be integrated at any time. There is only one system framework, however many modules are used. It contains the general components of the system, such as the measured data connection, the interface with the data management system (database), the auxiliary system, etc. On installation of further modules, it is only necessary to implement the module-specific components and to integrate them in the system framework. The system framework therefore provides the following advantages:

▷ Maximum standardization of the individual modules,

▷ Minimum effort for development of a new module,

▷ Simple expandability of an installed module configuration,

▷ Flexible, customer-specific configurability of the system.

The basic hardware platform are advanced workstations which are powerful enough to allow the installation of several modules on one computer and simple adaptation of the hardware infrastructure to the configuration of the diagnostics system. The advantages of the DIGEST system philosophy (modular structure, system framework) are integration, expansion and simple configuration. A stand-alone solution can not provide these advantages. However, every DIGEST module can be run in stand-alone mode sacrificing the advantages described above, whereby it is necessary to make savings on hardware and software. If there is any interest in a stand-alone solution, it is possible to make a separate application of a module. Subsequent conversion to an integrated solution would then entail additional costs.

5.5 Software Application

In DIGEST the various modules are very different in character; while some are more analytically oriented and thus use conventional algorithms, others are based more on the knowledge and experience of experts. These modules incorporate the latest software technologies such as expert systems, neuronal networks, genetic algorithms and fuzzy logic.

5.5.1 Expert Systems (XPS)

In those diagnostics modules in which deterministic solution procedures can be used for programming, the discrete steps involved in the solution of the problem in hand are sequentially processed in the form of algorithms, as in conventional computer programs. Figure 5.5.1.0 shows the different programming methods between programming on an algorithm basis and the programming structure of an expert system.

Conventional monitoring and analysis systems acquire the plant status using measured data that is numerically combined, analyzed and displayed. By comparing measured and characteristic variables with specified statuses, it is thus possible, for example, to detect deviations or limit value violations. The interpretation and combination of measured results is still a task for the operating personnel. In more complex cases, that is to say for comprehensive diagnostics tasks entailing, for instance, the combination of diagnoses from the module level, conventional data processing methods are too expensive and thus no longer applicable. Examples are the analysis and interpretation of frequency spectra such as are required for high-frequency measurement of electrical damage, shaft vibration or hydrogen sealing with typical fault sequences in the form of image comparison. Such complex problems can best be solved as “knowledge-based systems”, otherwise known as “Expert Systems”.

The programming structure of an expert system permits implementation of complex structures, and facilitates modification and extension of the software. These complex and demanding tasks are now accomplished by diagnostics systems, that thus make a considerable contribution toward relieving the work load on the operating personnel and reducing the sources of error in diagnostics. They extend the functions of analysis systems by the option of assessing the status of the plant, making predictions and recommending measures to be taken [9].

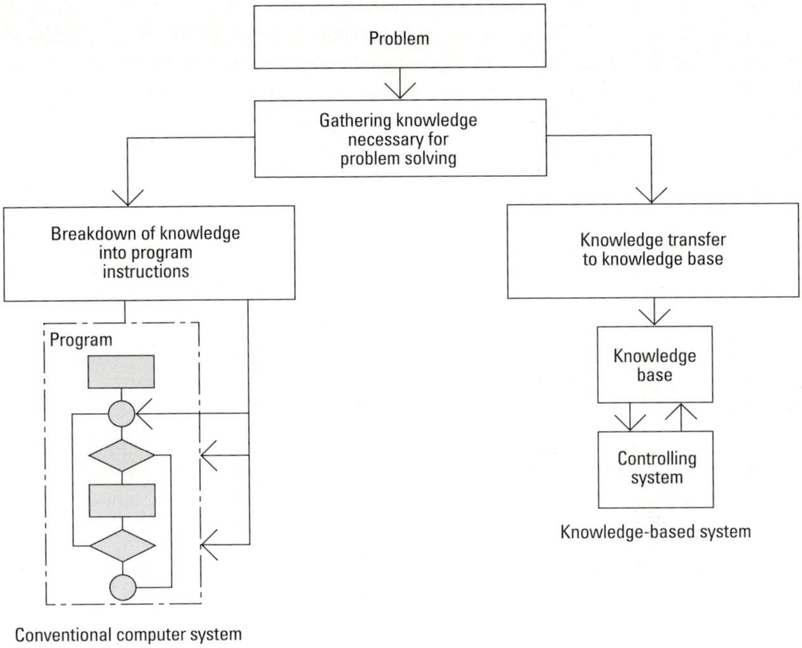

Figure 5.5.1.0 Program Structures

In expert systems, the correlations are no longer rigidly defined as is the case in conventional programming; instead the underlying knowledge is made available in explicit and declarative, and thus editable, configuration in a knowledge base. Usually, this is done in the form of (heuristic) rules derived from past experience. Such a system needs not to be complete from the very beginning but is usually built up gradually by testing, changing and adding to the knowledge base. This emulates the way experts work, as they also continue to learn throughout lives, revising and extending their knowledge on the basis of new experience.

The advantages of expert systems over conventional programming are simple maintenance, easy modification and expansion of the knowledge. Expert systems are therefore a suitable answer to the necessity for the flexible transmission of technological knowledge.

In the meantime, expert systems have become state of the art, with a large number of projects where these systems have been implemented successfully. There are nevertheless also expert system projects that have not proven effective, with the reasons for this often that the uncertainty inherent in the knowledge was not adequately represented. If a problem is characterized by uncertainties, the associated expert system must feature special mechanisms to accommodate this. A whole range of methods has been developed over the course of expert system history, with probabilistic models and so-called confidence factors gaining a certain degree of importance here. Recently, the use of fuzzy logic to solve such problems has become ever more popular. Expert systems based on fuzzy logic allow not only the implementation of expert knowledge, but also provide a method for the presentation of knowledge entailing uncertainties and for modeling uncertainties in the decision-making process [10].

Knowledge featuring uncertainty can occur in various forms, which can be characterized as imprecise definition, incompleteness and inaccuracy. Of importance here is the so-called linguistic uncertainty, which stems from the impreciseness of natural language.

Fuzzy Logic Expert System

The way expert systems work is likewise based on the ability to draw logical conclusions from given facts with the aid of the rules formulated in the knowledge base. In this context, classical logic assumes that a statement is either completely true or completely false and demands that all statements be classified accordingly. On this basis, a rule applies completely or not at all. This manner of representing and processing knowledge does not do justice to the way humans think. This is why expert systems have not been successful on a broad scale up to now.

It is the call for precision that makes it almost impossible to simulate human thinking by means of logic operations. Many human decision-making processes use concepts which have no precise definition. Most symbols used in human thinking have a fuzzy meaning. A characteristic feature of human thinking is that it predominantly uses "in-between" or imprecise concepts such as "quite", "hardly" or "lots of". A human expert does not formulate his knowledge in the form of:

"*If the temperature is higher than 640 °C and the pressure above 40 bar, then the control valve for the air supply is to be set to 0.4*",

but rather as follows:

"*If the temperature is very high and the pressure above normal, then the air supply should be cut back a bit*".

Fuzzy logic provides a mathematics-based method of expressing and processing this "fuzzy" knowledge. It enables the expert to describe a subject in his own familiar terminology, making use of the richness of natural language. He is relieved of having to apply a strict formalism that does not match the way he thinks or approaches a problem. Fuzzy logic has proven to be a powerful tool that also permits manipulation of imprecise, uncertain and incomplete knowledge.

In DIGEST, symptoms are described by tagging the various parameters and their manifestations (e.g. "temperature" may be "low"/"normal"/"high") with membership functions describing the degree of correspondence of a concrete reading to its manifestation class. For example, a temperature of 50 °C is translated into the description: “Temperature is high with an evaluation factor of 0.3”. These statements are linked by means of fuzzy operators. These not only take into account the AND, OR and NOT operators familiar with classical logic but also the fact that certain symptoms, if present, may reinforce (or refute) a putative diagnosis but if not present, are of no relevance to that diagnosis. The primary symptoms are linked with each other over several stages, at which interim diagnoses are performed, until a final diagnosis can be derived. The output of one rule may be introduced as a new symptom into another rule, that is to say the rules are interconnected in the form of a network. The system thus makes it possible to build up knowledge bases of any desired complexity.

The output from a diagnostics run is one or more diagnoses with different evaluations (degrees of confidence). An explanation is given for each diagnosis, together with an indication of possible actions to be taken. Figure 5.5.1.1 shows an overview of the various stages involved in the construction and the functioning of an expert system based on fuzzy logic. The files are allocated to the upper and lower levels shown in this Figure. The second level contains programs and functions. The files are created by the programs and updated appropriately (data base and symptom memory) or can be edited externally. This permits manipulation of the entire knowledge base via the editor. The diagnostics procedure is as follows:

▷ Further processing of measured values using the processing (PMCL) formulas and symptom generation on the basis of the stored symptom descriptions,

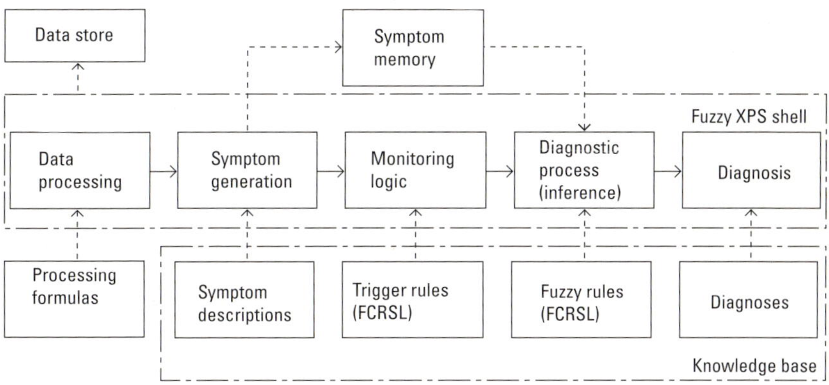

Figure 5.5.1.1 Structure of the Fuzzy Expert System

- ▷ Checking of diagnostics criteria by the monitoring logic, using trigger rules, Fuzzy rules (FCRSL),
- ▷ Processing of the rules contained in the rule base and generation of diagnoses with an associated confidence level.

On-Line and Off-Line Mode of the XPS

If diagnostics is to be wide-ranging and comprehensive, many process variables must be available. The on-line process variables are processed automatically by the XPS so that the diagnoses are made largely without user intervention. However, because the required measuring points are not always available, on-line diagnostics is complemented by off-line diagnostics in which the user can enter further process parameters or answer questions asked by the system. In this way, the automatically processed diagnoses are verified, refined or completed. If measuring points are added, which is possible at any time, the corresponding process variables are automatically processed in the on-line diagnostics. Figure 5.5.1.2 shows the structure of an expert system and the application in the field of steam turbine generator sets.

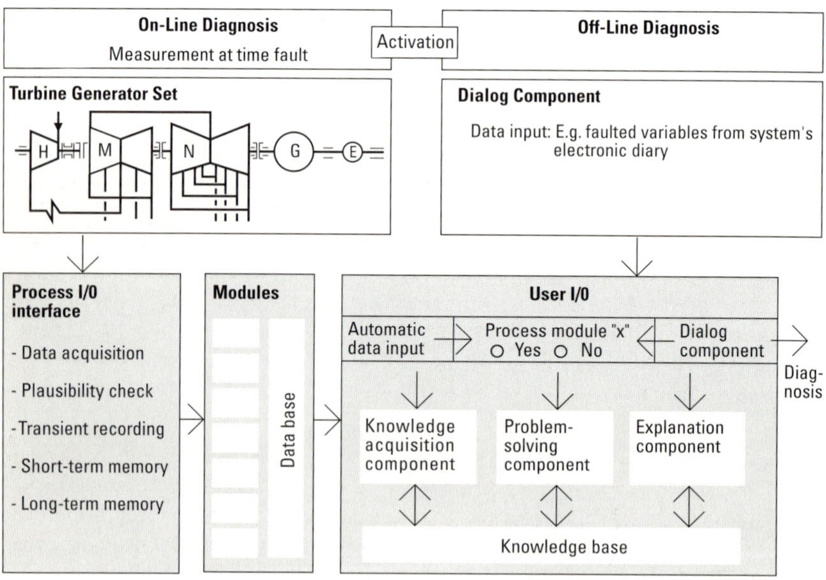

Figure 5.5.1.2 Steam Turbine Generator Diagnostics Concept

Integration of the XPS into DIGEST

Of course, the XPS is integrated into the general DIGEST system framework. Knowledge-based modules and the system-wide diagnostics therefore do not require any additional hardware but run with the other modules in a common standardized environment. The expert system calls, for example, the operator's attention unprompted to indicate an event (active approach). For this purpose, it projects the screen contents in a window on the monitor. By no means may the detection of an event be left to casual browsing. The operator receives the important data for assessment of the situation (malfunction) in a quick, well-founded and clear manner on the screen. For this purpose, the screen is subdivided into a number of windows which contain survey and plant diagrams with the possibility of zooming to view the high-lighted affected item, e.g. historical curves of the faulted process variables (any combination possible), plain texts describing causes and remedial actions for malfunctions and control diagrams of the affected control circuit. Examples of typical screen displays are shown in chapter 7 (DIGEST Modules in Application).

Because the diagnostics system can never be completely objective, maintenance and permanent updating of the knowledge base is essential. To test and experiment with the knowledge base, a simulation program can be provided that actively involves the customer in the experience gathering and optimization process. The knowledge base originally supplied by the system supplier can jointly be further developed with the customer to achieve progress in experience and know-how. The expert system thus ensures that state-of-the-art technological know-how is available.

5.5.2 UNIX Operating System

UNIX was chosen as the operating system for DIGEST because of the following reasons: UNIX-based operating systems are in widespread use and are available for many different hardware systems. The UNIX system's ability and stability to be ported from one hardware environment to another are what makes a large toolset possible. Widespread use in universities is one more reason for the continuing development of UNIX. Applications in engineering and computer science are already available and conversant in UNIX and C, which results in a huge increase in productivity. Because of the system's comfort established, experienced programmers are more willing to work on UNIX than proprietary systems.

Nowadays, all major system manufacturers offer UNIX in addition to their own operating system. This ensures constant expansion and further development of UNIX for the most varied computer architectures with standardized interfaces. UNIX can be used both on stand-alone computers and networks. The architecture of UNIX permits multi-user applications and multi-tasking which provide the basis for the rapidly increasing use of UNIX. The system can be used on CISC and RISC processors and multiprocessor systems. UNIX also provides the option of developing complex programs efficiently using simple tools with the use of command interpreters and other mechanisms. The file system permits simple and free hierarchical structuring and easy maintenance. Binding, stable, manufacturer-independent and verifiable interfaces exist for UNIX systems that provide a secure basis for long-life applications. UNIX also supports open communication. Many groups communicate via mail systems and public domain software, forcing it to expand constantly.

5.5.3 Neural Networks

Artificial neural networks emulate the basic characteristics of the human brain, such as massive parallelism and the capacity to process incomplete, unreliable and contradictory information without any claim to model the human brain in any way [11].

Artificial neural net models have been studied for many years in the hope of achieving human-like performance especially in the fields of speech and image recognition. There has been a recent resurgence in the field of artificial neural nets caused by new net technologies and algorithms, analog VLSI (Very Large Scale Integrated Circuit) implementation techniques, and the belief that massive parallelism is essential for high performance speed and image recognition. These models are composed of many nonlinear computational elements operating in parallel and arranged in patterns reminiscent of biological nervous systems. Computational elements or nodes are connected via weights that are typically adapted during use to improve performance.

Artificial neural net models or simply “neural nets“ go by many names such as connectionist models, parallel distributed processing models, and neuromorphic systems. Neural net models explore many competing hypotheses simultaneously, using massively parallel nets composed of many computational elements connected by links with variable weights. Instead of capturing complicated processes in formulas and then solving the problem analytically or numerically, neural networks can learn the model using only training data in the form of examples (data-driven modeling).

Neural net models are specified by the net topology, node characteristics, and training or learning rules. These rules specify an initial set of weights and indicate how weights should be adapted during use to improve performance. Both design procedures and training rules are the topic of much current research.

The art of designing neural nets is the art of minimizing the complexity and maximizing the ability to generalize. In training, it is necessary to divide the data material available for the application into a training data set, a validation data set and a test data set. The validation data set is used to check during training whether the generalization error decreases monotonically with the number of training data passes or whether it reaches a minimum before rising again. The correct approach is to analyze numeric data of the weights, too, if

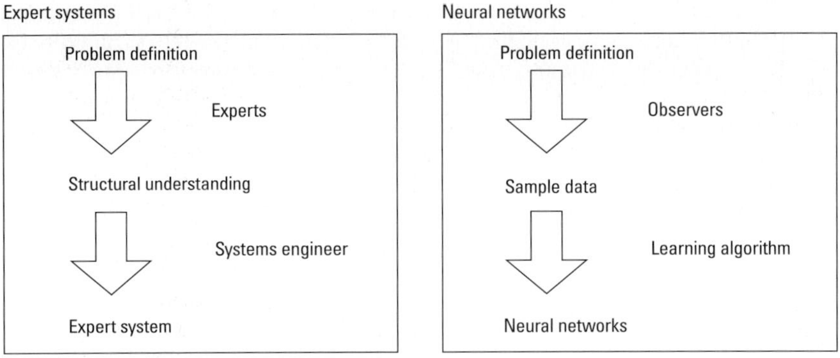

Figure 5.5.3.0 Problem Solution Expertsystems versus Neural Networks

minima occur in the generalization error. Those weights that reflect only background noise (determined using statistical methods) must then be eliminated and the training continued with a more rigid network topology. This is continued until all weights contain only relevant information.

As a universal approximation technique, neural networks are ideally suited for modeling, optimization, prediction and checking of multivariate nonlinear systems and processes. Many current closed-loop control and open-loop control problems (e.g. in robotics and process automation) are highly multivariate and nonlinear. They can often only incompletely be described by models. Neural networks are of great use for such processes because of their ability to learn from measured data. It is essential for successful processing of applications to integrate the problem to be processed into the network structure in an optimal fashion.

In the corporate department for Research and Development (ZFE) of Siemens in Munich and at Siemens Corporate Research (SCR) in Princeton, the high potential for use in applications of neural networks was recognized about a decade ago. Since then many applications have already been developed. The solution to a problem using a neural network, as can be found in technology-oriented tasks, is shown in Figure 5.5.3.0 next to a solution using an expert system for comparison.

5.5.4 Genetic Algorithm

The genetic algorithm method is a new technique for applications in the field of optimization processes involving a large number of parameters where a purely algorithmic approach does not bring the desired results and neuronal networks on the basis of learning programs for the sequences are not so well suited. The following is a brief description of the basic principles of the genetic algorithm function.

John Holland is the founder of the field of genetic algorithms [12]. To encode complicated structures and the power of simple transformations and to improve such structures, genetic algorithms are the problem-solving techniques with an astonishing property: They solve the problems by evolving solutions as nature does, rather than by looking for solutions in a more principled way. Genetic algorithms may serve as an optimization algorithm for a wide range of problem types. Some feature theory of evolution includes the features that Holland incorporated in his genetic algorithms:

- Evolution operates on encoding of biological entities, rather than on the entities themselves. Evolution takes place at the level of the chromosome.

The genetic algorithms operate on encoding of solutions to problems that must be decoded in order for them to be understood:

- Nature tends to make more copies of chromosomes that are more fit. This principle, sometimes called "survival of the fittest," says that the chromosomes in a population associated with the fittest individuals will, on average, reproduce more often than chromosomes associated with unfit individuals.

Fitness is determined by comparing an individual with the other members of its population. In a genetic algorithm, the user provides a function that evaluates chromosomes. This evaluation function plays the role of the environment for a genetic algorithm:

- Mutation and recombination occur in nature when reproduction takes place. Mutation operations introduce random diversity into the population. Recombination operations allow chromosomal material from different parents to be combined in a single child.
- Whatever nature "knows" about its search for good chromosomes is contained in the genetic makeup of those individuals that are currently alive. Similarly, whatever a genetic algorithm "knows" about its search for a good solution to the problems we pose it is contained in the genetic material currently in the population of chromosomes.

The genetic algorithm main process includes:

- An initial population of chromosomes is generated and each chromosome is evaluated with the user-supplied evaluation function.
- Reproduction occurs. Parents are selected from the population using a random selection procedure. Each parent's selection chances are biased so that the parents with the highest evaluations are most likely to reproduce. Children are made by copying the parents and the parents are returned to the population.
- Possible modification of children occurs. Mutation and recombination operators are applied to the children. Their chances of being applied and the points on the children at which they are applied are randomly determined.
- The children undergo evaluation by the user-supplied evaluation function.
- Room is made for the children by discarding members of the population of chromosomes. The worst population members should most likely be discarded. The children are then inserted into the population.

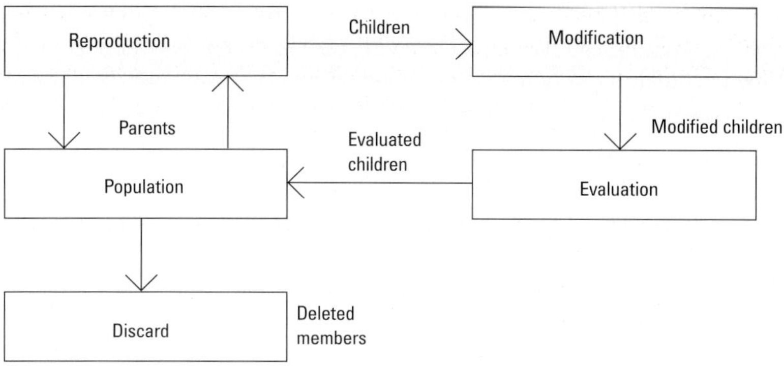

Figure 5.5.4.0 Procedure of Genetic Algorithm

The basic cycle is illustrated in Figure 5.5.4.0. The steps in this Figure are repeated until the algorithm is halted. The composition of the final population of chromosomes – the best chromosome(s) – is the solution of genetic algorithm to the problem. If all goes right, an initial population of relatively undistinguished solutions may develop the yield of the target algorithmic. In computers, simulation genetic algorithm represents everything by means of bit strings; a two-character alphabet is the simplest one when binary notation is employed. All evaluation functions return positive numbers, the higher the number, the better the chromosome.

Our in-house applications in the field of genetic algorithm include concepts for the utilization of extremely large data banks, fuel cycle planning for nuclear power plants and industrial projects in this field. This technology is being evaluated for solving process optimization in DIGEST (thermo-dynamic process optimization).

5.6 Software Features and Graphics

5.6.1 Software Features

The human operator's capacity to absorb information and the speed at which he can process information are limited. This applies in particular under unforeseen conditions (cognitive overload). Therefore the design of the highly interactive graphical operator environment serves to support the operator in the decision-making process. The critical factor in this respect is a display system which permits ergonomically optimized process representation using vector graphics, multi-window and user-friendly interaction features. Hence, DIGEST is an integral part of the existing control room design and process control philosophy. DIGEST was devised on the basis of important software and hardware criteria.

The system was:

- ▷ To use the existing infrastructure, such as transmitters, bus system and data acquisition systems, as far as possible, i.e. it was to be integrated as a service function into the process instrumentation and control,
- ▷ To be fully user-configurable on both the software and the hardware sides to satisfy the most varied requirements and conditions,
- ▷ To be based on international and widely used standards to ensure compatibility with various hardware platforms today and in the future, too,
- ▷ To exploit the progress made in computer technology, such as so-called RISC processors, and to implement modern software engineering methods which support knowledge-based methodology in addition to analytic algorithmic programming.

The X Window system environment (on UNIX basis) fulfills two essential functions: It forms the basis – by now likewise established as standard – for the graphic user interface and it permits the user program and the user interface to be distributed at will within a network (in a client/server architecture). This keeps the network transparent for the user: For example, he can call up and execute from so-called X-terminals all the modules installed in the network without having to know the structure of the network. Furthermore, any desired number of X-terminals can be connected up, permitting simultaneous observation and/or processing of several modules.

The requisite computer equipment complement can be configurated from case to case, depending on the client's wishes, the power requirements of the various modules and the instrumentation and control equipment present (e.g. data acquisition systems). Normally, up to five DIGEST modules can be installed on one computer. The system can then be flexibly extended stage by stage in line with the desired performance features. This means that it is neither necessary to implement right from the beginning a DIGEST system scope that might be envisaged only as a possible future option, nor to replace hardware to accommodate more sophisticated requirements not originally envisaged.

The user interface is implemented with the OSF/Motif tool kit based on the X Window system and suitable tool support. It permits simple and intuitive operation of the system even by a computer-illiterate user. Tool kits are libraries which provide the applications programmer with functions for generating and manipulating dialog elements. The tool kits afford simple and multiple-choice features such as menus, selection keys and tool boxes.

Push buttons and menus are used to initiate actions. Text input and display is by means of text labels, single- and multiple-line editable text fields and tables. Scroll bars and image displays are provided for the input of setting values. A so-called "style guide", an in-house code for the design of user interfaces, ensures a uniform, homogeneous appearance and standardized user control features across all modules. Programming is performed in

Table 5.6.1.0 System Basis

Hardware	DEC Alpha Station (200, 400, 1000) *
Operating system	DEC (UNIX)
Basic graphics system	XWindows
User Interface	OSF/Motif
Data Store	C-ISAM
Application Software	C(FORTRAN)
Communication	Ethernet, SINEC H1, AP, TCP/IP
Tools	Code Center UIM/X XMove Interleaf RCS

*) Example

language C and can also incorporate calculation programs already available in FORTRAN. Table 5.6.1.0 gives an overview of the most important tools and an example of a workstation in use for many applications.

Large systems intended for a long service life must operate continuously and being easy to master. Ease of maintenance and adaptability to changing requirements are also very important. For this reason, the implementation of DIGEST paid particular attention to the aspects of modularity of the software and object-oriented programming.

5.6.2 Graphics

Extensive instrumentation that acquires the condition of turbine generators is usually used for detailed process monitoring. Changes can be observed in trend plots. Dangerous conditions for important measuring points are indicated. The trend plots are useful tools for the interpretation of analyses and diagnoses. By zooming they permit easier recognition of small changes which are also displayed and stored in numerical form. Figure 5.6.2.0 shows a trend plot example for the various functions.

It is also possible to select individual events and correct them to match the various characteristics of the plant, i.e. display of trend curves for corrective measured variables in a clear, easily understood way. It is, of course, also possible to define the measured data to be displayed in the trends to provide operating personnel with a simple analysis tool. For clarification of complex interrelations between various measured data the XY graph is a useful tool. Instead of plotting the time at the X axis, any other data can be selected and

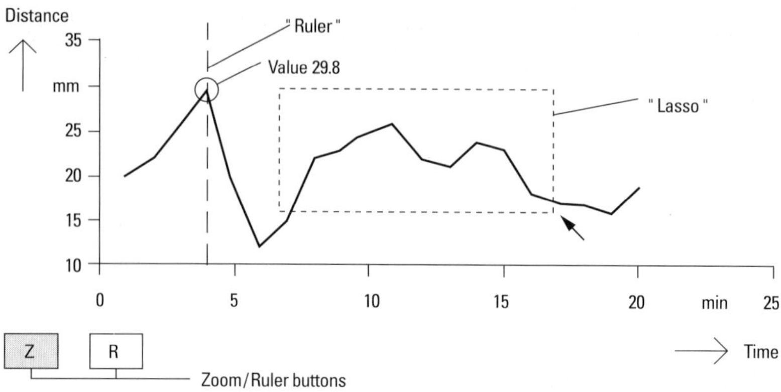

Figure 5.6.2.0 Trend Plot

plotted on this axis. Trend plotting of measured data is one way of using a modern tool to provide a better basis for analyses. This method is suitable for small turbine generators, for example, where the personnel perform analyses and where, for various reasons, it may not be viable to use large diagnostics systems.

Freely Configurable Graphic

DIGEST gathers and stores all data required for analysis and diagnosis. This permits further evaluation of these data at a later time. The graphics support capability of a workstation includes a whole range of different presentation features for measured values and derived (calculated) values. Special freely configurable graphics were developed to allow full use of these features, with the associated software suitable for use with all DIGEST modules.

These freely configurable graphics allow the stored data to be accessed and displayed in comfortable graphics. Features possible include post editing of the data (e.g. integration into curves) and overlaying of reference data. The display format implemented on screen can be in the form of a trend plot (values/time). XY plot (values/values) or as a polar plot (amplitude/phase), etc., as required. The measured data and characteristic parameters to be displayed on a diagram and the form in which this is to be implemented are set by configuration. The memory from which data to be displayed is read (e.g. daily memory, monthly memory) and the database and time range are likewise set by configuration. When configuration is performed, the associated diagram is displayed to allow this to be checked, with configuration not transferred into the relevant display until all settings have been made.

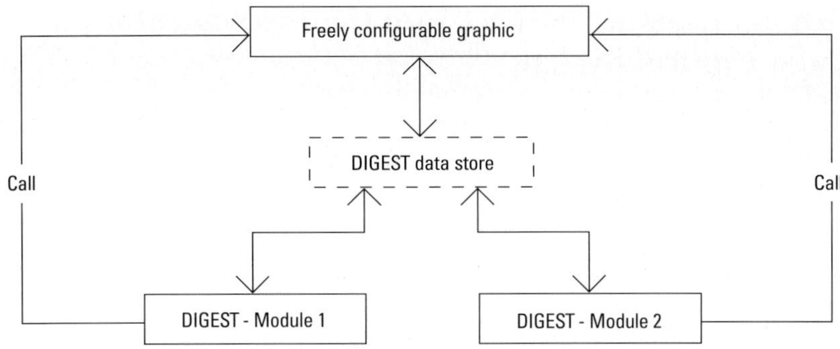

Figure 5.6.2.1 Concept of the Freely Configurable Graphic

The freely configurable graphics are used for a number of different DIGEST modules in addition to the module-specific displays provided in the associated standard user interfaces. The concept for this is shown in Figure 5.6.2.1.

6 Integration of DIGEST into the Instrumentation and Control System

6.1 Conceptual Features

In our plants, digital instrumentation and control technology are normally used. This allows full integration of DIGEST into the TELEPERM XP instrumentation and control system. Since the instrumentation and control system and DIGEST are based on the UNIX operating system, data, system, and diagnostics image material can be interchchanged and/or jointly used. The use of DIGEST in power stations with other I&C concepts, e. g. hard-wired I&C systems is possible. An integration or association of these systems depends, however, on the structure of the I&C concepts available [13].

The conceptual design of a diagnostics system associated with an instrumentation and control system is shown in Figure 6.1.0. All measurement signals already available from the instrumentation and control system that are required for the diagnostics system are read in via the plant bus. The scope of measurements required for the diagnostics system depends on the diagnostics modules implemented and in some module groups can be considerably greater than that required for standard on-line monitoring.

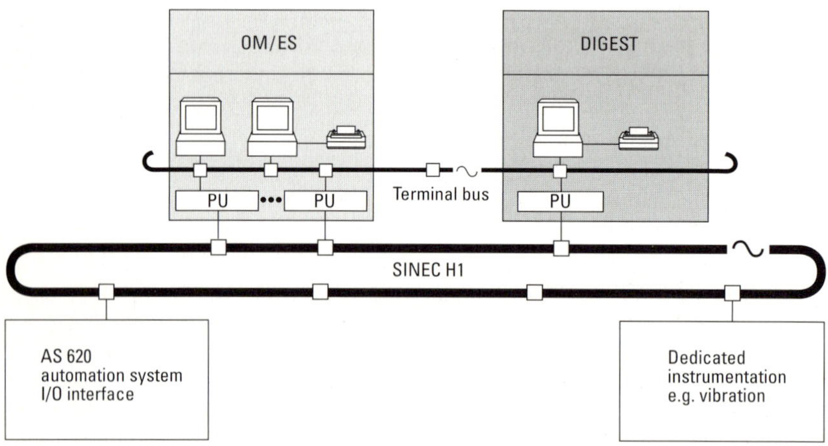

Figure 6.1.0 Integration of DIGEST in the I&C Concept

Specific modules may also require the use of additional sensors or mearsurement systems. Measurement signals that have to be backfitted for the diagnostics system are likewise acquired and conditoned in the normal instrumentation and control system and made available to the diagnostics system via the plant bus. Signals for specific modules are generated via the special-purpose instrumentation (e. g. by means of an automation unit in the SIMATIC series) and passed on to the analysis computer via a separate bus (SINEC H1); often it is also due to the high data load on the bus.

As a rule, the parameter measured for standard purposes is checked for compliance with its specified limits. For diagnostics purposes, however, the essential parameters are subject to a detailed analysis at the process interface. Of particular importance is a plausibility check which is performed in the preparatory phase to avoid incorrect diagnoses being made on the basis of unrealistic parameter readings. Provision is also made for derivation of a parameter value on the basis of other parameters if its own dedicated sensor should fail. This ensures that only plausible values actually representative of the process are used as input for diagnostics. In many cases, this requirement calls for mathematical models to recognize and define equilibrium conditions, gradients and fluctuations.

6.2 Software Aspects and Hardware

The results output by DIGEST can be displayed at the X-terminal. Conditioned data are stored in the diagnostics computer, either in long-term or short-term memory depending on the type of data in question. The data are available in logs and diagrams but can also be called up on screen. All messages, alarms and recommendations generated can be stored by the system and recorded in an “electronic diary”. The data stored in the diagnostics system can be stored by fault event and exported via disks or other data carriers. Fault-related data files may also be transmitted via telephone modem. This makes it possible for the turbine manufacturer to perform detailed fault analysis on his own computer. Equipment is provided for graphic processing of faults, with comprehensive, detailed, and polychromatic analysis of the diagnosis-triggering events. A hard copy print-out of each display can be generated.

The type of computers required will depend on the scope of diagnostics modules specified. High-performance workstations or PC’s with a high storage capacity are used. As a rule, several modules can be installed on the same computer. Even at the time of installation of any module, provision is made for the complete system framework, permitting the system to be extended by adding on further DIGEST modules at any time. This computer has the entire user interface on its screen. If the capacity of a computer is exhausted by the installation of further modules, one or more computer system units, depending on the scope of the software required, can be added.

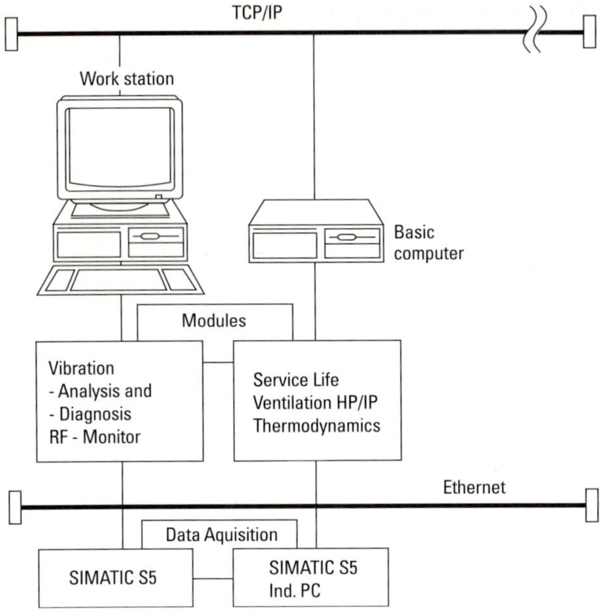

Figure 6.2.0 Computer Configuration of DIGEST Modules, Example

Figure 6.2.0 shows a conceptual design in which several DIGEST modules and the requisite computers for a typical application of DIGEST modules are used.

High-performance workstations have proven suitable for applications to date. The diagnostics system may also be made available to another user in plant operations management or the interconnected grid system operations management center. The associated terminals can be integrated into the operator control and monitoring level of the main control room hardware. The number of terminals required will depend on the specified scope of diagnostics.

7 DIGEST Modules in Application

In this chapter an overview is given of the diagnostics modules developed applicable for the various types of steam turbines, gas turbines and turbogenerators. The modules are described as follows in terms of their task, functional principle and software conception. The most important modules for the turbines and turbogenerators diagnostics are currently available from the DIGEST family, and most of these modules have been service-proven and cost-effective.

7.1 Vibration Analysis and Diagnosis Module

The dynamic response of a turbine generator unit is determined by its system properties. These properties are significantly influenced by the rotor stiffness and masses, the relationship between natural frequencies and the service frequency (tuning) and the bearing characteristics (damping, stiffness). Besides, the vibration behavior of a turbine generator unit depends on many influences which originate from operating procedures and operational disturbances that cannot be fully eliminated. The magnitude of the allowable vibration due to the residual imbalance and the range of vibration change during plant operation must be matched to design conditions to ensure reliable operation and safe start-up and shut-down of the turbine generator unit [14].

Vibration monitoring of turbomachinery by measurement of the absolute bearing housing and relative, respectively absolute shaft vibration, that allows evaluation of the vibration behavior based on guidelines, has proven effective in the past. The disadvantage of this monitoring concept, however, is that the information of frequency spectra and the vector angle and magnitude as integral parts of the measured signal are not used for a detailed assessment of vibration behavior. Experience has shown that certain changes in vibration behavior of turbine generator units and malfunctions were not recognized early enough in individual cases.

The use of suitable analyses procedures which calculate characteristics for assessment of vibration behavior and which keep many years of professional experience stored in the knowledge bases of diagnostics systems now permit reliable assessment of incipient disturbances and the avoidance of malfunctions to enhance the safety and reliability of turbine generator units. Maintenance work on the turbine generator unit can be planned in good time to eliminate the causes of identified potential malfunctions. In combination with other analysis and diagnosis modules, this module is a prerequisite for changing over from time-based maintenance inspection schedules to condition-oriented maintenance. Condition-oriented maintenance is cost-effective.

To be able to detect the changes in the condition of turbomachinery, however, an analysis system is needed which can process the large numbers of fault symptoms detected into extensive graphic displays. As a rule, such a system is provided as a supplement to the vibration monitoring and protection functions of the on-line monitoring system of the turbine generator unit. A diagnostics system can be installed in addition to the analysis system. That system aims to interpret the fault symptoms in an expert fashion and faces a sophisticated task. It has only recently become possible to solve these tasks thanks to a high standard of development in instrumentation technology, to today's powerful computers with large memory capacities and the advances in software technology. Figure 7.1.0 shows an overview of the structure of the system for vibration analysis and survey diagnostics.

Task of the Module

The task of the vibration analysis module is to detect the change in vibration behavior of the turbine generator unit at a certain time and under defined

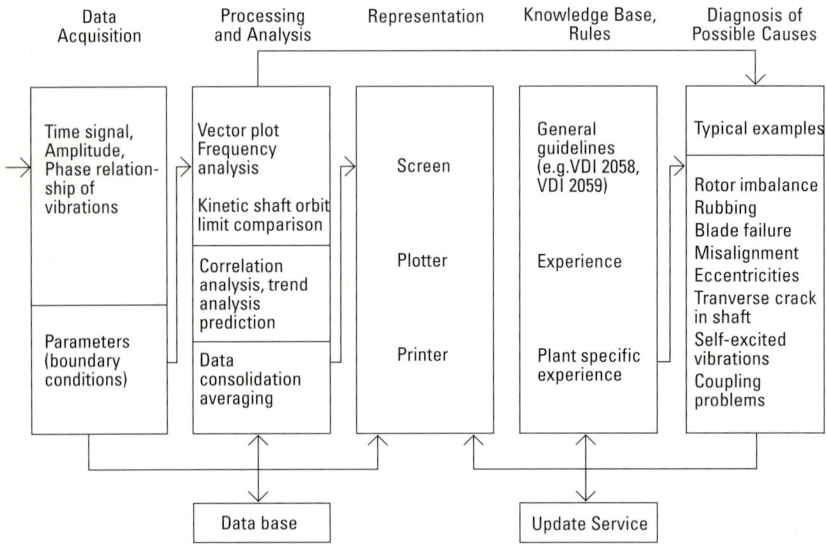

Figure 7.1.0 Vibration Analysis and Survey Diagnostics

operating conditions and to process the measurement signals into characteristics which make comprehensive analyses possible.

An essential basis for assessment of the vibration behavior is an continuos comparison of the actual vibration behavior with the stored "normal data" which represent the undisturbed vibration behavior. For this purpose, analysis involves processing of the measurement signals and derived characteristics in the given time and frequency range. This entails transforming the primary (measured) vibration signals into secondary information (e.g. frequency spectra, vector diagrams, etc.). Sophisticated graphs and time histories are generated which make it possible to perform a diagnosis on the basis of the characteristics integrated in the graphs and of their change over time. Certain measured values and characteristics above and beyond the defined thresholds are also recorded.

7.1.1 Analysis Concept

Measured Data Acquisition

Figure 7.1.1.0 shows the basic arrangement of the vibration pickups and their position on the shafts and bearing housings of a turbine generator unit. One measurement plane at each end of the shaft and at least one plane between two bearing housings are generally used as measuring locations. The anisotropic nature of the bearings results in non-circular kinetic shaft paths; generally ellipses but straight lines in the limiting case. In order to fully register shaft motion (S_{max} in accordance with VDI 2059/2), it is necessary to use two displacement pickup sensors arranged at a 90 degree offset. This also applies for measurements on the bearing housings. Thus two separate measurement planes with two pickup sensors each at a 90 degree offset are provided for each bearing housing of the steam turbine generator set for registration of the absolute bearing housing and relative shaft vibration.

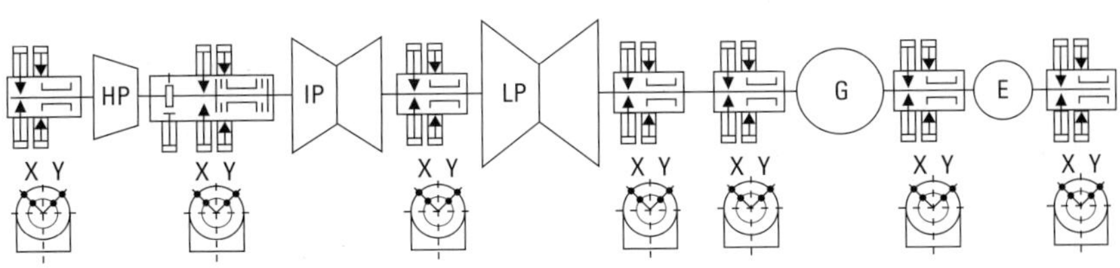

Absolute bearing vibration

Relative shaft vibration

Key phasor

Figure 7.1.1.0 Basic Arrangement of Vibration Pickups and their Position

From these measurements the absolute shaft vibration in *X*- and *Y*- direction is calculated and finally combined to form the resultant absolute shaft vibration. Assessment of vibration behavior is based on absolute shaft vibration. Absolute shaft vibration is the appropriate method for monitoring large steam turbines because bearing housing vibration reach such high levels that they must be considered in the evaluation of vibration behavior.

Apart from this, for further processing, one pulse per revolution is required to provide information on the phase angle of every periodic component in addition to the vibration amplitudes. The pulse generator (key phasor) is usually located in the front bearing housing. In addition to the vibration characteristics, the main process variables are also acquired by the analysis system.

The relative shaft vibration is measured by means of non-conducting, proximity-type pickups (eddy current principle). The absolute bearing vibration is measured by means of piezoelectric accelerometers. The same transmitters are used for signal processing for on-line monitoring. Protection criteria for high vibration trip are derived from the bearing housing vibration on the basis of 2-out-of-2 logic. For this reason there is an additional vibration pickup provided at the x axes at each bearing housing. If the protection system logic is based on 2-out-of-3, an additional transmitter is required.

Measured Data Conditioning and Analysis

The logical and physical structure of the system provides for a division into separate signal preparation (conditioning) and analysis stages (Figure 7.1.1.1). The signal conditioning computer communicates directly with the power plant process, it fulfills real-time requirements. Data acquisition and preparation are performed by the conditioning computer. For a thorough analysis it is necessary to transform the measured primary vibration signals (time-related signals) into secondary information (characteristics). The conditioning computer performs the calculation of the characteristics on the basis of the scanning intervals in phase with the rotary shaft motion and transmits data to the analysis computer. The essential data conditioning is as follows:

- ▷ Transformation of the measured primary vibration signals into frequency spectra (by means of the fourier analysis) and determination of the phase position and magnitude,
- ▷ Converting velocity and acceleration signals into displacement signals,
- ▷ Mathematical generation of the absolute shaft vibration from the absolute bearing housing vibration and relative shaft vibration,

▷ Calculation of S_{max} and other vibration characteristics in accordance with VDI Guideline 2059/2 as follows:

S_{max}	Maximum shaft deflection (resultant, both measuring directions),
Offset	Zero-frequency quantity, shaft position,
S_{pp}	Vibration signal amplitude (peak to peak),
S_{ges}	Vibration amplitude sum (in one direction of measurement),
S_{ω}	Amplitude of vibration signal after filtering to 1_{ω},
φ_{ω}	Position of phase after filtering to 1_{ω},
$S_{2\omega}$	Amplitude of vibration signal after filtering to 2_{ω},
$\varphi_{2\omega}$	Position of phase after filtering to 2_{ω},
$S_{3\omega}$	Amplitude of vibration signal after filtering to 3_{ω},
$\varphi_{3\omega}$	Position of phase after filtering to 3_{ω},
S_{n1}	Amplitude (sum) in the first non-harmonic frequency range,
φ_{n1}	Position of phase of S_{n1},
S_{n2}	Amplitude (sum) in the second non-harmonic frequency range,
φ_{n2}	Position of phase of S_{n2},
S_{n3}	Amplitude (sum) in the third non-harmonic frequency range,
φ_{n3}	Position of phase of S_{n3}.

The vibration characteristics S_{n1-3} represent frequency spectras which can be parameterized by the operator. The vibration characteristics are calculated from the vibration signals at one-second intervals and transferred to the analysis computer.

A time sequence of digitized values for each measurement plane in two mutually perpendicular measuring directions is then available in the computer. The scan rate and conditioning intervals for measured data depend on the current operating status. Besides the vibration signals, further the vibration behavior influencing boundary conditions are also captured by the

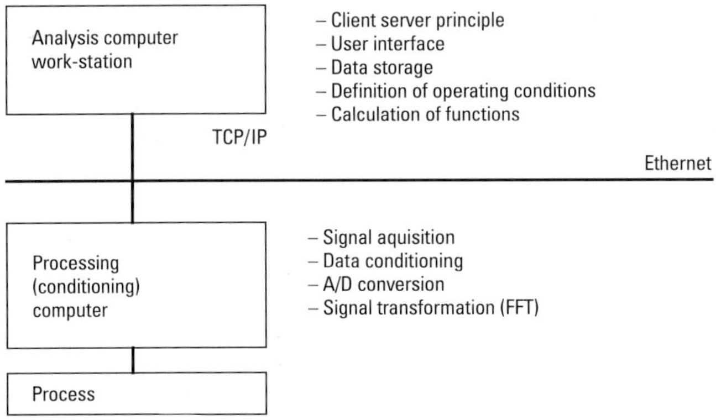

Figure 7.1.1.1 Data Manipulation and Hardware Concept

system. A radio-controlled clock ensures precise timing for synchronization with the plant computer.

The data will be stored in the analysis computer depending on the plants operating condition and changes of the process variables and are, however, available for further analyses. The analyses run on-line, interactive and full-time. The characteristics calculated are represented in graphic form on the analysis computer. These also give access to the stored data. Data capture delivers measurement signals from the signal conditioning computer; these signals may contain errors. Any errors occurring are recorded. If an error is detected in any characteristic, the entire data record associated with the measuring point concerned is marked.

The operator interface of the analysis computer is separate from the function module so that the operator can switch it on and off at any time. The operator can view the ongoing analyses at any time, modify it, or extract the information in which he is interested in a suitable form. The analysis computer is the main interface between the user and the conditioning of the vibration signals. It controls the signal conditioning computer. The computer incorporates a watchdog function which in case of a fault regenerates the necessary processes.

The Principle of Vibration Analysis

The vibration behavior of a line of shafting is also influenced by various operating conditions of the power plant. For this reason it is necessary, in response to a change in the vibration behavior, to determine whether the change has gone beyond the "normal behavior" range. Thresholds are only defined for conditions that constitute normal operating states. No thresholds are defined for non-steady-state conditions or turning gear operation.

Influencing Variables

The normal vibration behavior is characterized by a vibration signature affected by known influencing variables, within the normal operating frame of the plant. The most important influencing variables are:

- ▷ Speed,
- ▷ Active power output,
- ▷ Condenser vacuum (backpressure),
- ▷ Excitation current,
- ▷ Heating steam extraction from heating steam turbine generator units.

The registration and dispatch of the influencing variables to the analysis computer are performed at one-second intervals. The system affords the possibility storing the thresholds for up to 10 influencing variables associated with various normal operating states in order to be able to explain the resulting changes in the vibration characteristics or to assess their influence.

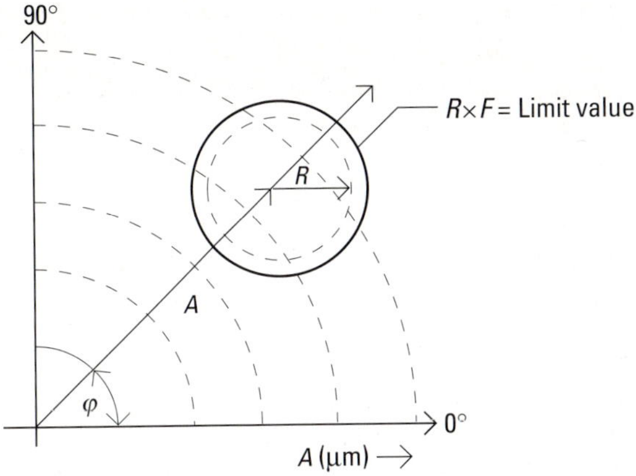

Figure 7.1.1.2
Definition of Limit Values

The normal vibration behavior is established on the basis of empirical values. Allowable scatter bands are defined at the time of commissioning, as a rule from service-related fingerprints at the commencement of the turbine generator units service life. The adjustment of these thresholds is automatically performed by programmed and event-controlled data inputs.

Reference values for the vibration spectra are determined for the most significant service conditions of the plant (normal operating states). From the data defining the normal state, a normal vector and a tolerance circle, within which actual vectors may lie without a fault begin reported, are defined for each measuring point (Figure 7.1.1.2). Thresholds are established for each measuring point and for each defined frequency band. The amplitude A and phase angle (φ) for the normal state are entered, as is a radius R, within which the amplitude and the phase angle must lie. Multiplication by a factor F then yields the threshold for the normal operating state, which must not be exceeded. The vibration vectors are checked for deviation from the learned or entered standard operation vectors; if the limits are violated an alarm is issued or a diagnostics run is triggered. Changes in the vibration behavior within scatter bands are allowable. It is not necessary to restrict operation in any way, and the turbine generator unit can operate throughout the full agreed output spectrum.

The vibration behavior is further influenced by binary and boundary variables within the normal operation range of the plant.

Binary Variables

Binary variables are, e.g. circuit breaker status and valve positions. Operating experience shows that steam turbine generator units generally react to changes in these parameters with changes in vibration behavior which would not make it necessary to check for incipient malfunctions. The analysis system is capable to process up to 32 binary variables for positive determination of the operating condition.

Boundary Variables

Further to these binary variables there are, of course, a number of cycle parameters (boundary variables), which also exercise a certain influence on the vibration behavior of the turbine generator unit (e.g. steam temperature, casing temperature, bearing temperature, steam pressure, generator-side cold-gas temperature, etc.). These are referred to as boundary variables. The analysis system can process up to about 100 boundary variables.

The influencing, binary and boundary variables are, as a rule, fed in from the on-line instrumentation. The registration and processing of the binary variables are performed at one-second intervals. With the boundary variables required at ten-second intervals, the influence on the condition of the turbine generator set is sufficient recognized.

Operating Status

A further distinction is made with respect to non steady-state operating conditions between the initial condition of the turbine generator unit and changing physical parameters. The system distinguishes between the following various operating states:

Steady-State Operation

The turbine operates at rated speed. Steady-state operation begins when the speed lies within the defined scatter band for steady-state operation (parametrizable) and the load exceeds a certain limit. Acquisition of the time-related vibration signals is synchronized with the rotor speed. The scan frequency is adjustable within a wide range at the time of commencement of the first operation. For example, with a parametrization for rated speed as follows: Scan frequency 1600 Hz and a scan time (time window) of 2.56 s.

The scan frequency 1600 Hz, together with the scan time t = 2.56 s, produces a resolution of 0.4 Hz and a measuring range of 0 to 800 Hz. The time-related signals are transferred to the analysis computer at two-hour intervals or on request.

Normal Operation

This is a special type of steady-state operation. The turbine operates at the rated speed and has been in one of the defined normal operating states for more than 30 minutes. Normal operating states are determined by the influencing variables. The system is capable of defining 50 different normal operating conditions.

Non-Steady-State Operation (start-up and shut-down sequences)

Non-steady-state operation is in force when the speed lies between rated speed and turbine turning-speed. The type of sequence (cold start, warm start, hot start, shut-down; measured data for shaft temperature are essential input here) is determined by the initial conditions. Limits are not monitored during non-steady-state operation. The measured data are stored for later analysis. Time-related signals are acquired st for $\Delta n = 120$ (3000 rpm) or $\Delta n = 60$ (1500 rpm) and transferred from the conditioning computer to the analysis computer. Characteristic values are calculated at intervals of $\Delta n = 15$ or at lower speeds (< 750 rpm) $\Delta n = 7$ rpm.

Turning Gear Operation

Plant-specific upper and lower speed values define the envelope for turning gear operation. During turning gear operation, acquisition of the characteristic, influencing, binary and boundary variables continues. They are stored at five-minute intervals (parametrizable).

Standstill

Measured data acquisition stops when the speed lies below the lower limit for turning gear operation. The analysis system determines the operating state of the turbine automatically from the measured variables.

Storage Concept

With the great amount of vibration signals, calculated vibration characteristics and the dependence on influencing, binary and boundary variables captured during operation and the various operating conditions of the turbine generator unit, the threshold of data storage would be easily exceeded. In order to assure the storage of the essential data a convenient data storage concept is imperative. The following concept for data compression is used:

- ▷ Data averaging (determining of time-related mean values and storage of this data),
- ▷ Event-oriented data storage of data depending on the following criteria:
  - – Change of normal operation conditions,
  - – Violation of thresholds (change of vibration behavior by influencing variables or vibration characteristics),
  - – Storage of specific, parameterizable data for various reasons, if storage is not performed otherwise.

The essential vibration data of the system will be stored. The data storage concept and the characteristic groups of data stored are depicted in Figure 7.1.1.3. This Figure represents an example; the concept is adaptable to the various demands of power plants.

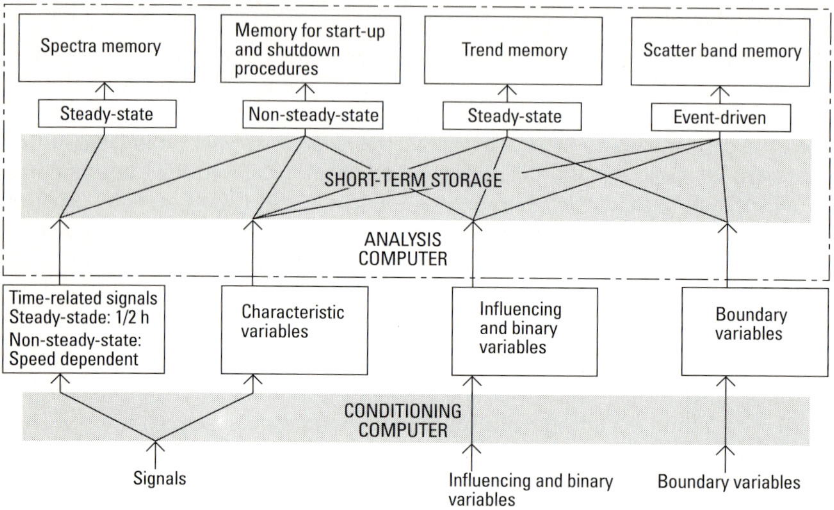

Figure 7.1.1.3 Vibration Analysis Storage Concept

The different types of memories applied in the vibration analysis system are:

- Storage of non-steady-state data (start-up, shut-down procedures). Storage of vibration characteristics, influencing, binary and boundary variables and time-related vibration signals (depending on speed) for two hours in the short-term storage and then in the memory for non-steady-state data. With a maximum of 30 sequences for start-up and shut-down.
- Trend memory, the stored variables are compressed as follows:
  - ▷ The daily memory contains five-minute mean values,
  - ▷ The monthly memory contains one-hour mean values,
  - ▷ The annual memory contains daily mean values.
- Scatterband memory. Long-term storage of vibration characteristics and the influencing, binary and boundary variables are stored here in the event that measured data lies outside the specific scatterband. The number of the data records is limited to 20,000 after which the oldest data records are overwritten (ring buffer).
- Time-related signal memory (spectra memory). During non-steady-state operation, the time-related signals are transferred into short-term storage at two-hour intervals. Subsequent storage is then in the form of spectra for a maximum of 40 days.
- Early crack detection data storage (see chapter "*Visualization*").

The time-related data records and the memory content are parameterizable. Long-term data storage is possible on DAT (Digital Audio Tape) or MOD (Magneto-Optical Disk).

Visualization

The graphics performed of a powerful workstation offers various possibilities for displaying measurement signals and characteristics. Standard graphics and freely configurable graphics are provided for visualization of the various data. Standard graphics are normally provided for visualization of often required analyses. This type of graphic is available by the menu of the analysis computer without any operational procedure by the operator. Violation of standard operating limits that lead to activation of the diagnostics and violation of alarm annunciation limits are indicated in the standard graphics by a change in color. The freely configurable graphics enable data to be selected for visualization from the data storage, particularly for in-depth analysis.

Standard Graphics

The following displays are provided by the standard graphics:

Signal Overview

Plant status display for all signals (measuring points). The signal overview forms the main window and gives an overview of the overall plant status. The respective measuring point is characterized by a specific color that defines the condition of the signals, i.e. the display shows where the bearing housing and shaft vibration limit thresholds may have been exceeded.

Bar Chart

Display of the peak values (S_{max}, S_{ges}, S_{ω}, $S_{2\omega}$, $S_{3\omega}$) for all signals and the influencing variables. The peak values are displayed as a bar chart with a table of values shown underneath. Bar charts are provided for display of analysis in real time or from stored data records for any point in time (freely configurable graphics). The bar charts comprise the colors green for variables within the allowable value range (no annunciation), yellow for variables in violation of the allowable value range (warning threshold exceeded, e.g. diagnostics run initiated) and red for violations of the respective limit (alarm threshold exceeded). Defects are shown in blue. Gray indicates no monitoring function is operative.

Vector Overview

Display of the peak values for all signals and the influencing variables underneath. Peak values are displayed as a vector diagram with the threshold for diagnosis and annunciation. The status of the measuring point is displayed via indicator flags in the table of values. A selection between the display of shaft and/or bearing house vibration is possible. The vector display is possible on the basis of the 1st, 2nd or 3rd harmonic speed. The ruler function is also available for the vector diagram. In addition, the definition of a reference value is possible that allows to display the difference to the actual values.

Trend Functions

Display of up to 8 measured variables versus time from short-term storage. The trend functions show the behavior of selected vibration characteristics during the last minutes. The display features a scaled marker function to permit identification of the precise values of the variables at the point at which the scaled marker stands. A zoom function is also provided (see chapter 5.6.). Of course, the individual influencing variables and boundary variables can also be displayed in combination with any selection of vibration parameters. The selection of reference data for comparison with actual data is also possible by using the freely configurable graphic.

Start-Up/Shut-Down

Presentation of the most recent, or selected (reference values), start-up or shut-down sequence as a Bode diagram. The measured peak data for vibration (shaft and/or bearing housing) with phase angle as a function of the speed are displayed for comparison with the data of the last operating state. Displaying these vibration characteristics versus speed, shows shaft resonances and the critical rotor speed. Conditions at which an increased vibration occurs will be recognized. A number of displays can be provided for representing these turbine generator unit operating modes.

For further analysis of start-up and shut-down procedures, there is the possibility to display the spectra of the vibration signals over these procedures in the form of a so-called cascade diagram. Speed-dependent spectra changes of the vibration signals of the shaft and the bearing housing in each direction can be compared during transient conditions. It is also possible, e.g. at steady-state condition, to compare the spectras determined at different time. The three-dimensional configurated graphic of the cascade diagram is allowed to be turned spatial. These displays illustrate the start-up and shut-down transients especially in the critically phase and reveal any changes in the start-up and shut-down vibration behavior.

Vectors and Spectra

Display of the peak data and the spectra of a measurement plane in steady-state operation. The peak data are displayed as a vector diagram with a table of values shown underneath, with the spectra shown in a separate X/Y-presentation format. The status of the measuring point is displayed via indicator flags in the table of values. From the measured data with the time frame of two hours, the mean power components of a signal based on frequency spectrum will be calculated. As carried out in other displays, the amplitudes of the bearing housing and/or shaft and the threshold limits can be represented in the vector diagrams. The frequency range of the spectras is selectable by the operator on the basis of 150 or 400 Hz.

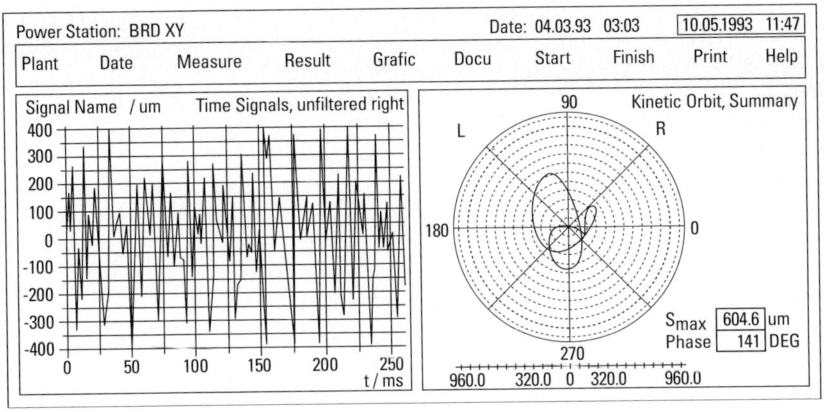

Figure 7.1.1.4 Measuring Results, Kinetic Orbit and Unfiltered Signals

Time Functions and Orbit

The graphic features allow the presentation of the time-related signals (e.g. 1024 measuring points) and the kinetic shaft path (orbit) in a measurement plane. In this display, the amplitude is represented by the distance of the orbit from the origin of the coordinate. The position of the vibration point is represented by the phase angle. Figure 7.1.1.4 (derived from a hardcopy) shows an example of the recording of the locus curve (shaft kinetic orbit) together with the unfiltered time signals. A vector display of the amplitudes and phase angles of vibration variables make it possible to recognize changes, e.g. resonances, via the vector quantity. The changes can also be indicated in numerical terms with the aid of a scaled marker. It is possible to select the time-related signals of the shaft and/or bearing houses or the respective kinetic shaft path. The system allows also the separate representation of the time signals or the difference of these signals.

Eccentricity, Steady-State Shaft Position

The eccentricity of the shaft is represented in an attractive manner. The steady-state shaft position can be determined by means of the non-contact pickups. This involves calculating the location of the shaft centerline as a function of speed on the basis of the bearing geometries and clearances.

Early Crack Detection

The objective of this function is to detect with a high degree of certainty a transverse crack in a shaft early enough for the turbine generator unit to be run-down to turning gear operation or standstill before damage occurs [15]. An automatic detection procedure is implemented in the analysis system. The most important sign of rotor cracking at constant operating speed is the progressive change in vibration amplitudes as the crack grows. Due to the superposition of vibrations caused by rotor imbalance, the discernment of crack influences is dependent on actual rotor balance conditions. In addition, during coast-down a vibration spike at half-critical speed due to the double-frequency excitation component would indicate the presence of a crack. The detection threshold is a crack depth of approximately 25 – 30 % in a turbine generator shaft. Basis of this function is an algorithm which has been developed by Siemens/KWU. The specific data for early crack detection is stored in a separate subpart of the memory.

Freely Configurable Graphics

The freely configurable graphics enable data to be selected for display from various classes of data (non-steady-state data, daily, monthly or annual memory, scatter band memory or turning gear operation). A time range or time point can be selected. The axes can be linear, logarithmic, percentage (only Y-axis) or inverse. Various functions for data (smoothing, differentiation, integration, difference between two variables) are also available. Data can be displayed, e.g. as a trend diagram, X/Y plots, spectra, orbits, cascade diagrams, bar charts, coherence diagrams, auto and cross correlation diagrams. There is a specific criteria available within the freely configurable graphic to select data groups from the memory. The diagrams of the graphics can be stored and recalled under their address.

Records

To give an overview of the process data, records of parameters, operating conditions, daily records and message sequence records can be generated. Provision is also made for listing of parameters and data in table form. With a tool available up to 8 signals measured can be "freely" configured.

7.1.2 Diagnostics Concept

Diagnosis is the automatic evaluation of data and analysis results by compilation and association for fault recognition. The purpose of vibration diagnosis is to recognize incipient changes in vibration behavior with a view to defining faults. The operating personnel is informed about the condition of the turbine in verbal messages, and eye-catching graphic displays are provided to illustrate the results of diagnosis. A diagnosis is initiated via monitoring and analysis by the vibration module if the pre-defined thresholds for the pertinent data and characteristics are exceeded. If a number of DIGEST modules are linked with a higher-level diagnosis, violation of the vibration parameter thresholds triggers not only a diagnosis by the vibration module but also the analysis and diagnosis by other modules provided in the system (Figure 7.1.2.0). Diagnosis should augment shaft vibration monitoring with other modules. The high validity of the vibration analysis is attained by thermodynamic analysis and would be supported by the bearing module and analysis of the overall shaft alignment.

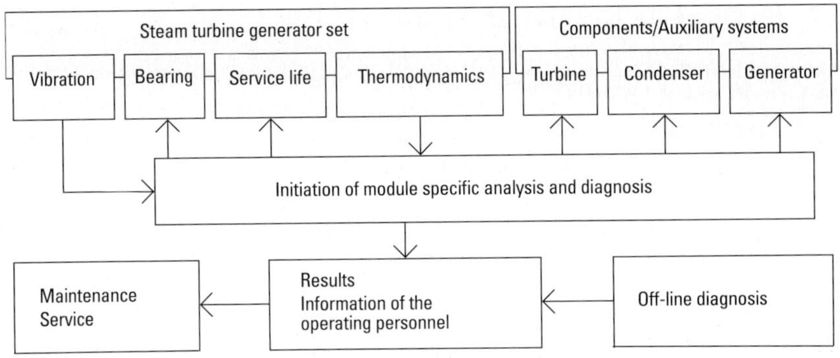

Figure 7.1.2.0 Diagnostics Function of Steam Turbine Generators

A vibration diagnostics system would employ the fuzzy logic expert system developed for DIGEST. On-line diagnosis is backed up in interactive dialog by results gained from off-line measurements. The basis of the diagnosis is the expert knowledge supplied by the turbine manufacturer. This expert knowledge is represented in the form of a matrix, the rules for which are stored in the expert system. The rules for definition of the faults are stored in the knowledge base of the diagnostics system. The expert knowledge is based on experience gained from information feedback on faults that have occurred in the operating utility's plants and/or in plants built by other suppliers. It is in the scope of service that Siemens/KWU would regularly update and add new rules. The vibration diagnostics system is able to interpret the malfunctions on the basis of the associated symptoms. An overview of significant malfunctions recognized by the vibration module is given as follows:

- ▷ Rotating blade break,
- ▷ Faulty bearing pedestal mounting,
- ▷ Incipient shaft crack (included in the analysis),
- ▷ Shaft distortion (line of shafting),
- ▷ Oil whip, oil whirl,
- ▷ Thermal turbine shaft distortion (rubbing of shaft seals),
- ▷ Disturbed bearings (e.g. restricted axial expansion, tilted bearing),
- ▷ Turbine foundation plate distortion,
- ▷ Shaft eccentricity (turning gear operation),
- ▷ Thermal generator shaft distortion (e.g. asymmetric cooling of generator rotor),
- ▷ Rotor imbalance due to different reasons,
- ▷ Loose shrink fit (e.g. couplings, disks).

7.2 Service Life Evaluation Module

The assessment of thermal stresses and the collation of information on the fatigue sustained by stressed components is an important task in connection with ongoing efforts to extend the service live of power plants and their components. The service life module group is already familiar under the name of "turbine stress controller" (chapter 4.6.3). It traces and evaluates the thermal load cycles that cause low-cycle fatigue and creep fatigue of representative parts such as turbine casings and rotors. It also traces the associated stress cycles in the rotor centerline with regard to crack growth. The cumulative usage factors and the usage fractions calculated give a timely indication of the need to inspect individual components and also reveal when the theoretical critical time for the onset of cracking has been reached.

These modules also afford opportunities for improving and optimizing operations management. Continuous "on-line monitoring" of fatigue effects makes it possible, within certain limits, to control variable thermal stressing and to come to an optimal adaptation to the desired mode of operation of the plant. The modules required will depend on the diagnostics targets. For instance, the "creep" module will not be required for saturated steam turbines since the temperature of the components is not relevant for creep fatigue. Particularly for large steam turbine generator sets and in case of certain grid conditions (short-circuit and short-circuit trip) torsional stresses of the shafts of the turbine generator can be taken under consideration with an additional submodule.

Hardware Configuration

The hardware configuration of the turbine stress controller is described under chapter 4.6.3. In case of using the service life module as part of DIGEST, the analysis of the service life consumption of the steam turbine is being represented at a workstation.

7.3 Thermodynamics Modules

7.3.1 Thermodynamics Modules for Fossil-Fired Power Plants

The thermal efficiency of power plants and the effective terminal output of steam turbine generators are dictated by a number of influences. Deviations from the expected status (design data) are often attributable to a variety of causes, so that it is impossible to analyze the changed process accurately and to adopt the requisite counter measures without computer assistance. The thermodynamics modules detect deviations of heat rate and power output from their standard values as computed by water/steam cycle analysis on the basis of plant-specific thermodynamic data [8]. The momentary status can be identified via continuous or cyclic recording of the operating parameters and calculation of derived data. Using continuously performed analysis, the thermodynamics module calculates the heat rate deficits even of those parameters requiring extensive calculation of their effect on the heat rate. The process data can be used to identify the status of the components within the water/steam cycle.

On-Line Module

Various types of on-line diagnostics may be performed for example, simple global efficiency analysis, or comprehensive diagnosis of the entire water/ steam cycle and its various components (options are shown in chapter 5.4, Table 5.4.0).

On-Line Module with Off-Line Calculation Program for Water/Steam Cycle

As experience with power plant instrumentation shows, older plants may not have the instrumentation points required in order to perform a comprehensive thermodynamic analysis of the process. Whereas the required instrumentation points can be implemented at relatively low cost into plants in the planning phase, backfitting instrumentation for plants already in service is expensive. In such plants, it is possible to perform global efficiency monitoring on-line on the basis of the available instrumentation points (major measuring points only) and using the thermodynamic computation program separately to analyze the process changes or for optimization purposes with changed process parameters. The input of these data that are not available on-line, is manual.

Off-Line Module

An off-line module is available performing thermodynamic calculations as applicated for the above on-line module. This module operates as a simulation program to perform computations of thermodynamics based on power plant parameters and configurations for the various plant conditions. The input of data is, of course, manual.

Task of the Modules

Even relatively small deviations from design parameters can result in a considerable loss of efficiency. Changes in the design conditions are assessed and correlated with the various influencing factors to permit detailed information to be gained on the causes of discrepancies in thermal efficiency and/or terminal power output. As a matter of course, the diagnostics system concept also includes storing the data in compact form for subsequent analysis at any time, for trend displays, or for comparison with data from other operating phases. The function of the analysis module is shown in Figure 7.3.1.0.

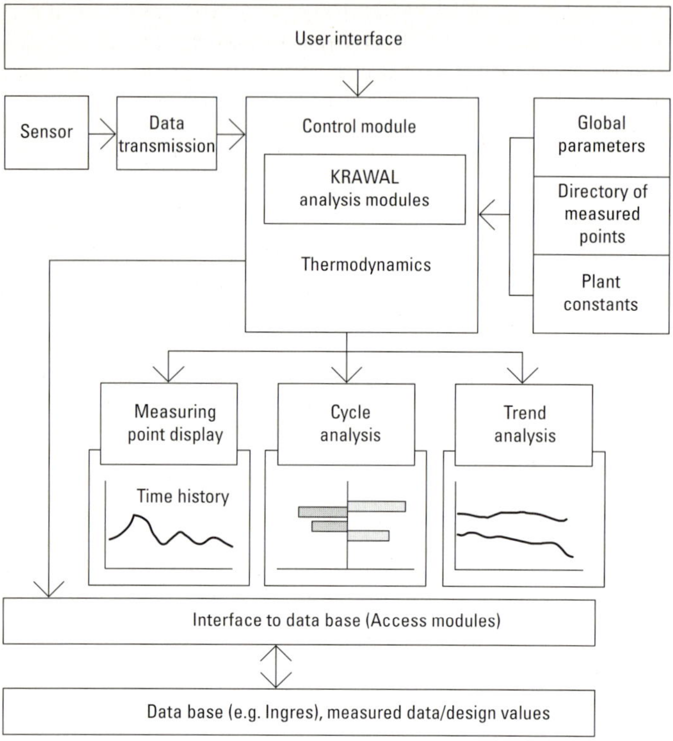

Figure 7.3.1.0 Thermodynamic Analysis

Thermodynamic Computation Program for the Water/Steam Cycle

Structure

The thermodynamic computation program (KRAWAL) is the basis for the computation of the water/steam cycle of power plants. The program calculates the (warrantable) data of the water/steam cycle (excluding the steam generator) and calculates essential thermodynamic data of the key components of the water/steam cycle. It can also be used for optimization of power plant components, subsystems, and to perform design calculations and computations for service conditions that deviate from the design conditions.

Efficiency Computations of the Steam Turbine

The overall expansion efficiency is calculated using the efficiencies of the individual turbine stages. The efficiency of each stage is calculated by means of the individual loss method according to “Traupel” (Thermische

Turbomaschinen, Thermodynamisch-Strömungstechnische Berechnung). The profile constants are derived from measurements. The electrical output of the turbine is calculated with the aid of these expansion points and the steam flows through each stage. The electrical output at the generator terminals is computed allowing for the mechanical losses of the turbine generator set and the losses in the generator. Once the expansion curves in the turbine stages have been determined, computations can be performed for the entire cycle.

Depending on the requirements of the case, the steam conditions in the turbine and in the cycle components can be calculated for a power plant of fixed design, or variable optimized design. If the turbine geometry is known, then the expansion efficiencies can be calculated as a function of load, which can then be used to compute precise data for the expansion curve at various part loads. The part load calculations also determine the pressure losses in all vessels and piping of the cycle and the terminal temperature difference of the heat exchangers. To sum it up, the analyses and calculations for the off-line and on-line modules offer the following essential options:

- ▷ Cycle computation based on design data,
- ▷ Cycle computation based on specified data,
- ▷ Evaluation of influencing factors on the basis of deviations from specified data,
- ▷ Computation of part loads for specified data,
- ▷ Computation of altered configuration, e.g. one feed water heater closed down.

Factors influencing efficiency which are the result of slight changes in requirements or in the operating mode can thus be readily analyzed. Potential for optimization can be quickly recognized. The data obtained from the simulation computer can be superimposed on a plant overview display. Figure 7.3.1.1 (derived from a hardcopy) shows an example of the scope of calculations performed.

Presentation of the Computation Results

The results of the computations are stored in a data file and can also be displayed on screen. The printout consists of separate data records for the cycle computation data and the steam turbine data. The printout shows the essential mass flows, the electrical power outputs, and the characteristics of the power plant. Expansion curves for the turbines and the variables of state and mass flows for all pipes in the power plant can be shown. The overall thermodynamic data of the turbine stages, feed heaters, the condenser and the pumps are also indicated. Use of the computation program permits graphic display of the results of the cycle computation according to various aspects (Figure 7.3.1.2, derived from a hardcopy).

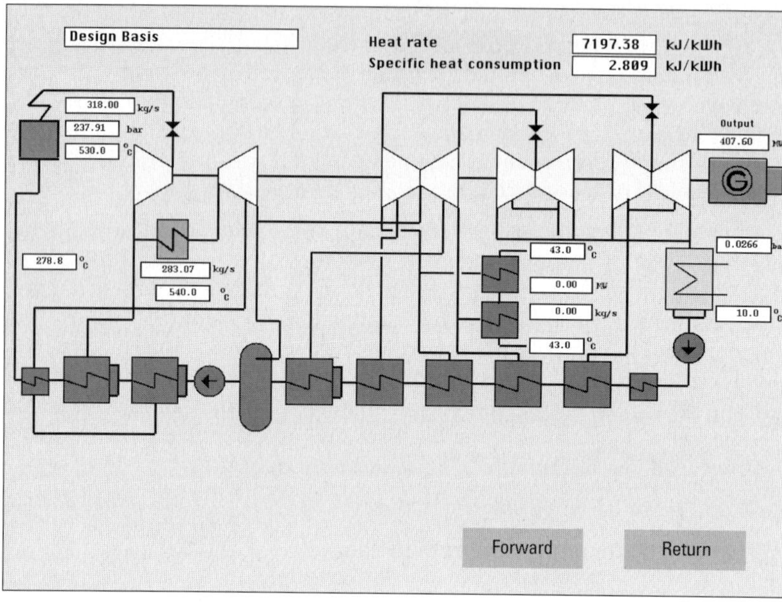

Figure 7.3.1.1 Plant Overview Display

Various means of visualization show, for example, the influence of changed data in graphic form or provide a direct comparison of different operating modes in terms of heat rate and output. Load conditions for which computations are performed can be stored, e.g. for use as a starting base to calculate the effects of parameter changes or to permit rapid consultation on new conditions. Different operating modes, obtained by changes to the feed water heating line, for example, are compared with each other in terms of their effects.

Data Acquisition and Configuration

The measurement signals already present in the instrumentation and control system are read into the analysis computer via the plant bus. In addition to the signals already present, it may be necessary to provide new instrumentation points. Instrumentation required for thermodynamic monitoring of power plants depends on the scope of the thermodynamic analysis. Great attention is paid to performing plausibility checks on the measured signals. These checks are performed at several levels to ensure the physical consistency of the input data. When the data is read in, the values are checked to ensure that they lie within a given range. The input check (filter) provided for this purpose is open to modification. All variables are allotted a quality tag.

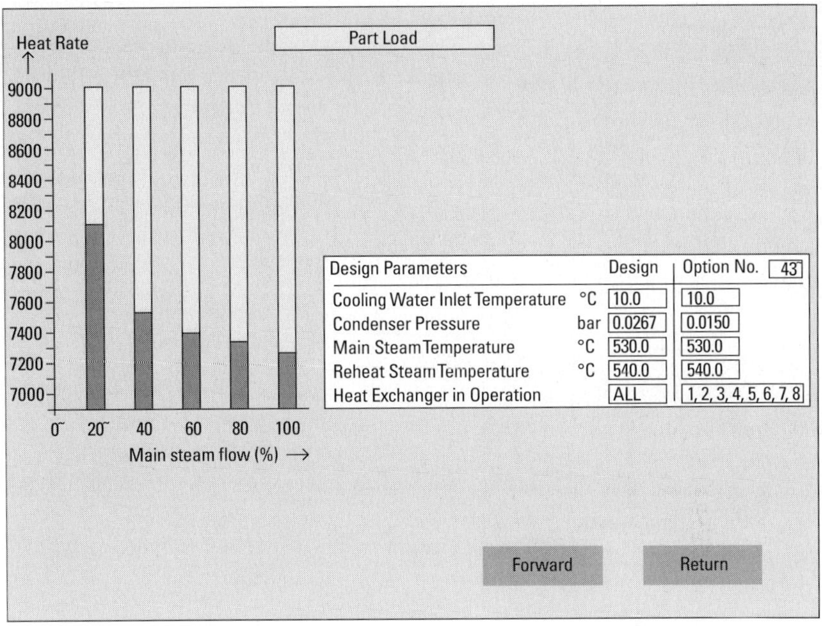

Figure 7.3.1.2 Heat Rate Relative to Design Condition for Part Loads

7.3.2 Thermodynamics Module for Combined-Cycle Power Plants

The high energy efficiency of combined-cycle power plants [16] with their complex heat flows has led to an increase in complexity of the instrumentation and control systems with the result that control room personnel are confronted with information processing problems. Consequently, the thermodynamic analysis for support of process control are of key importance and are in use along-side with process diagnostics. The purpose of such systems must therefore be to considerably enhance the information offered to the plant crew and to provide an overview of the actual status of the plant performance. It is practically impossible to continually calculate this process manually. Deviations from the optimum point of the operating mode are often attributable to complex causes. It is therefore impossible without an analysis system to assure early detection of any deterioration to the heat rate and to analyze these accurately in order to initiate appropriate counter measures [17].

Figure 7.3.2.0 shows the heat flow diagram of a combined-cycle power plant with triple-pressure cycle with steam reheating, incorporating gas turbines whose exhaust heat drives a steam turbine via heat recovery steam generators.

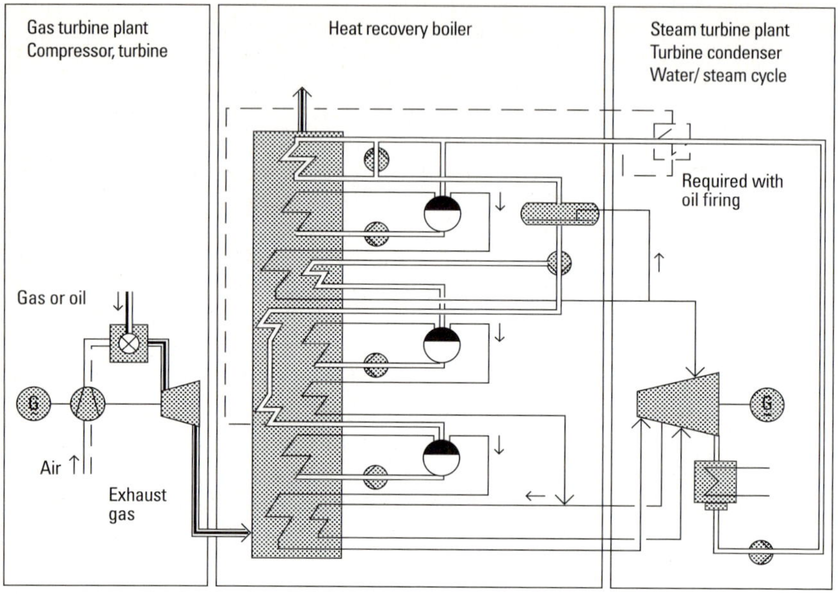

Figure 7.3.2.0 Combined-Cycle Power Plant, Cycle Arrangement

The modules required for the thermodynamic analysis are the following:

▷ Gas turbine (GT),

▷ Heat recovery steam generator (HRSG) and

▷ Steam turbine and water/steam cycle (ST,W/S-C).

Structure of the System

The thermodynamic gas turbine module for the combined-cycle power plant is equipped with a separate computer for each gas turbine generator. In addition to on-line gas turbine monitoring and diagnostics, general tasks are handled for the module in use. The heat recovery steam generator submodule performs thermodynamic computations to determine steam production in the individual heat exchanger sections. The basis for this is the actual plant operating condition with the momentarily available heat in the gas turbine exhaust, the water and steam-side parameters, and the heat exchanger surface areas, which can then be compared with the results of the actual measurements. The steam turbine module continuously monitors and performs the thermodynamic calculation of the steam turbine and of the water/ steam cycle. The thermodynamic computation program KRAWAL is the basis for this procedure. The various modules perform the following tasks: Acquisition and storage of measurement signals as a function of operating status. Analysis of deviations from reference data. Calculation of steam and gas turbine output (incl. compressor) characteristic data and calculation of the power plant efficiency and heat rate.

The calculations require steady-state conditions of the plant. The computations are coordinated by a control module, which also services the interfaces between the individual modules and the user interface. The control module also manages the central data base and coordinates the calculation assemblies of the various modules (Figure 7.3.1.0) for calculating the individual load points. As soon as steady-state operation has been achieved, the gas turbine module starts the thermodynamic calculations for the gas turbine. This module then transfers the values to the calculation module for the heat recovery steam generator and the steam turbine unit to permit computation of the reference values for the water/steam cycle, which are based on the actual output of the gas turbine. Finally, all values required for computation of the output of the overall power plant are transferred to the corresponding calculation module.

Trend analysis is intended as an additional diagnostics tool for the user to assess the technical state of the plant. Such a trend analysis provides a time history of a variable with respect to other variables. Any deviation of certain variables from set limits are annunciated to the operator and give indications of incipient problems. The selection of limits is open to modification.

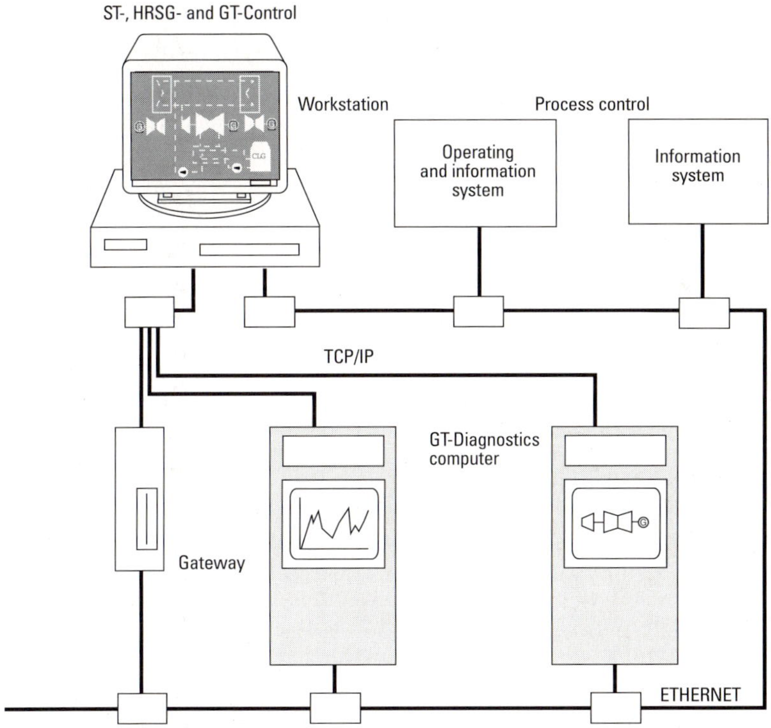

Figure 7.3.2.1 Diagnostics Concept Combined Cycle Power Plant, Hardware

Hardware

The hardware configuration of applicated modules is shown in Figure 7.3.2.1. The system consists of one workstation which performs the thermodynamic calculation of the plant performance and for the thermodynamic calculation for the steam turbine, water steam cycle and heat recovery boilers. Each gas turbine PC is responsible for monitoring the condition and performance of one gas turbine. The PC's are connected to the workstation via a TCP/IP bus. Signal acquisition of gas turbine data is performed via the plant bus (Ethernet).

Measured and calculated data from the gas turbine are sent to the workstation via the bus at regular intervals. The workstation is linked to the plant bus. The data is conditioned by the I&C system and transmitted at regular intervals to the workstation calculation processes. The use of separate computers for the gas turbines has advantages for stand-alone operation of the gas turbines and permits a number of analyses of mechanical operating performance in addition to the thermodynamic computations, if desired. The workstation can be replaced by a Pentium PC.

If calculation was initiated automatically, measured values are written to the process values data base to continue long-term recording. Calculated values are set to zero in the data base. The user has access to the process value buffer to read the data.

Printouts of messages and charts are done on a matrix printer. Hardcopies of the workstation monitors are sent to a color graphic printer.

7.3.3 Unit Commitment Module

In the past, determining the cheapest production policy of power plant units was usually left to the power plant operating personnel, who used a largely empirical approach for this task. Such automated systems for determining the reference variables that have existed up to now are based on a linearization of the functional correlation between the electrical, district heating, and/or process steam output of the power plant unit and the thermal power input into the power plant unit. However, due to this linearization, the computation procedures run through for the various systems giving imprecise results [18]. Furthermore, by employing a simple mathematical model for this functional correlation, it is, as a rule, not possible to take into account an ongoing plant operating condition, such as the loss of a heat exchanger or of a condenser half, in the determination of the optimal production policy.

To achieve the highest cost-effectiveness in operation of the entire plant, distribution of the demands among the power plant units has to be optimized. This task is handled by the DIGEST module "unit commitment".

Task of the Module

The task of the unit commitment module is to find the optimal production policy taking into account the plant status, i.e. the ongoing cycle configuration. Optimization is made with respect to fuel costs as well as start-up and shut-down costs of the plant units. In order to minimize the fuel costs, the demanded load has to be optimally distributed among several power plant units according to economic criteria on the basis of thermodynamic computations. These computations take into account the instantaneous status of the individual units and of the parameters which impose certain limitations in terms of loading and heat extraction.

Further to optimizing the load distribution among several different power plant units, it is also expedient to optimize units of the same type and rating this way, since at part load or in the event of abnormal operating parameters or faults, for example loss of feed water heater(s) or condenser tube bundles, it may be desirable for economic reasons to change the relative shares of electrical power, heating steam extraction and process steam extraction in the overall output of the power plant unit. The unit commitment module performs its computations for the entire electrical power of the units to be optimized, up to the point at which the power plant units feed into the public supply system, that is to say, it makes no allowance for power transmission losses or bottlenecks in the power distribution system.

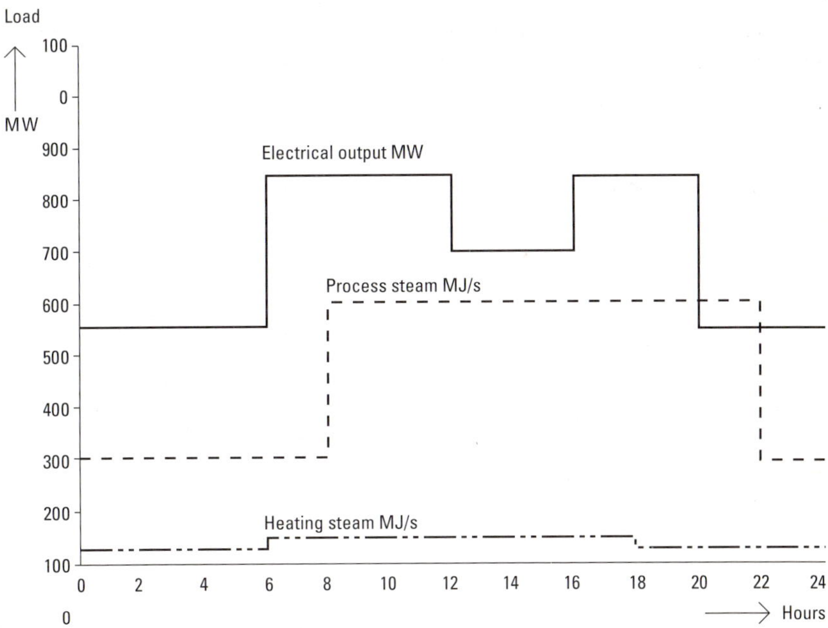

Figure 7.3.3.0 Unit Commitment, Load Cycling Schedule

The Principles of Unit Commitment

Unit commitment takes into account the following inputs and plant conditions:

- ▷ The load cycling schedule; Figure 7.3.3.0 is typical for a power plant in Germany,
- ▷ Priorities with respect to load cycling and heat extraction,
- ▷ Heat extraction for heating and process purposes,
- ▷ Current specified steam and load data with their upward and downward limitations, ambient conditions (e.g. cooling water temperature),
- ▷ Current plant configuration (e.g. allowing for shut-down feedwater heaters or condenser tube bundles),
- ▷ Unit run-up and run-down times,
- ▷ Cost of purchasing power from external sources,
- ▷ Heat extraction via auxiliary boilers.

The subjects of optimization are the selection of units with their actual condition and systems to be operated and the distribution of loads, since an optimal production policy may call for one or other unit to be shut-down. In the case of power plants consisting of several units, especially during the night, it may be more cost-effective to shut down one or more units than to operate all units at part load. It may also be more economical to run individual units requiring a high servicing effort at full load rather than operating some or all units at part load.

Optimization must also take into account the losses that occur when units are started up again, as the unit does not start feeding power into the public supply system until after it has been run-up and synchronized. For this reason, the program makes allowance for start-up costs in the unit commitment computations. Unit commitment is performed on the basis of the load cycling schedule, which records the predicted demands such that time is discreted into, e.g. half-hourly intervals. Provision has been made for commitment of up to 10 power plant units of different designs operating in the same interconnected power grid. The basic task of unit commitment is shown in Figure 7.3.3.1. Unit commitment is a complex optimization task because:

- ▷ Binary and real parameters have to be optimized at the same time,
- ▷ There exist constraints between unit outputs within the same time slice (load balance equations),
- ▷ There exist "cost coupling" different time slices (start-up costs),
- ▷ Technical conditions affecting the achievable load points have to be taken into account (heat and power cogeneration).

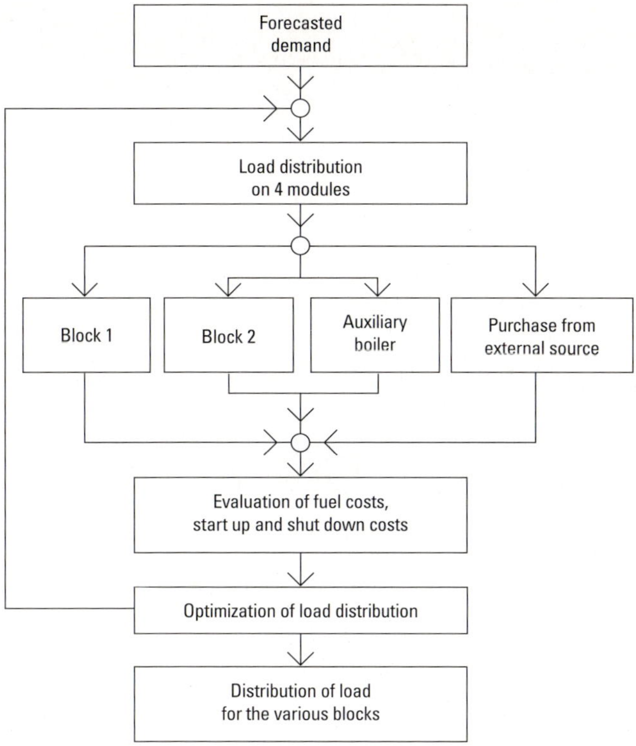

Figure 7.3.3.1 Unit Commitment, Basic Task for a Power Plant with 2 Blocks

Fuel costs cannot be accurately modeled by means of a simple (e.g. linearized) mathematical model, so such an approach cannot achieve the desired cost-saving effects. In order to calculate the power generation costs correctly a thermodynamic computation program has to be employed which performs precise calculations, such as the one we use (KRAWAL, chapter 7.3.1). Our concept of unit commitment employs modern methods of optimization (genetic algorithms) and interpolation (neural networks) and thus yields high-quality solutions.

Optimization Strategy

The optimization strategy consists of three stages. An overview is given in Figure 7.3.3.2.

1st Stage (Genetic Algorithms)

An optimization procedure by means of "genetic algorithms" (chapter 5.5.4) is used which, inspired by evolution in nature, is suited to solve complex

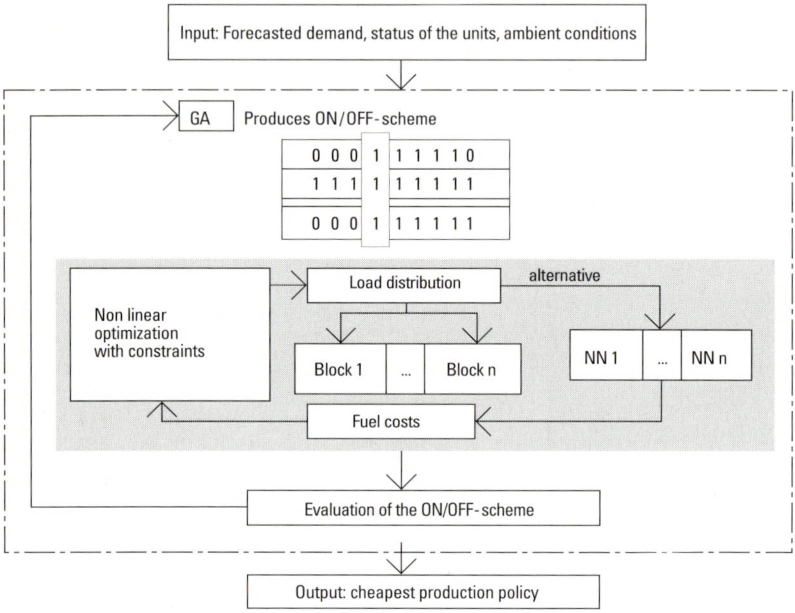

Figure 7.3.3.2 Unit Commitment, Optimization Strategy

optimization problems. This approach is specially suitable for global optimizations with discrete and continuous parameters. A genetic algorithm makes up the top level of the optimization strategy and generates possible ON/OFF scenarios for unit selection. That is to say, the genetic algorithm operates only on the binary variables that indicate at what times which unit is active. The unit selection is evaluated on the basis of the start-up and generation costs over the entire planning horizon. The most cost-effective allowable operating scenarios win through. In order to improve performance, the genetic algorithm acts on building blocks, i.e. highly-fit partial ON/OFF-scenarios.

2nd Stage (Optimization of Load Distribution with Constraints)

For the units selected to be in operation, a lower-level optimization algorithm determines the most cost-effective load distribution for a given time interval. The optimum distribution is calculated independently for each time slice. The load distribution is evaluated according to the sum of generation costs caused by each individual unit in generating its allotted share of the total output. The result is the most cost-effective load distribution obtainable. A program for optimization under constraints (Sequential Quadratic Programming or SQP) is used to determine the optimum load distribution for a given unit selection.

This program gets as input all parameters specifying load demand, status of the generating units and ambient conditions. Intermediate results can be logged and displayed. To enhance computation speed, the SQP optimization is performed with the aid of a rapid-calculation equivalent cost function. Then the cost calculation is critically reviewed and, if necessary, adapted by the thermodynamic computation program KRAWAL.

3rd Stage (Output of Optimum Load Conditions)

The optimum load conditions are calculated by means of the thermodynamic computation program (KRAWAL). In this third stage, the thermal input required within the individual unit to generate the assigned share of the total load is calculated by simulation. This may involve investigating various operating modes of the unit to determine which gives the optimum values for the required load point. If the requirements cannot be fulfilled, the thermal input for the most similar load points achievable is calculated. Thus KRAWAL is used to simulate, optimize as necessary, and determine the nearest achievable load point. Infeasibility of a required load point is taken into account as a violation of a constraint in the optimization of the load distribution.

Acceleration of KRAWAL-Simulation by Neural Networks

Each evaluation of a scenario proposed by a genetic algorithm involves calculating an optimum load distribution for all time slices. Computing each optimal load distribution by exact KRAWAL-simulation is very time consuming. Therefore, to accelerate optimization, KRAWAL is replaced in the early phases of SQP optimization by neural networks (see Section 5.5.3) by way of approximation. To achieve a constellation appropriate to the tasks involved in optimization, a number of harmonized neural networks are used for the specific tasks covered under the relatively broad scope of the computation program. Simulations have shown that the functional correlations elevated by KRAWAL are monotonous and continuously differentiable within the ranges in question. It thus follows that:

- ▷ KRAWAL can be quite accurately approximated by neural networks,
- ▷ Optimization for load distribution within individual time slices requires just a few steps,
- ▷ Local minima in the optimization of load distribution present no serious problem,
- ▷ The optimums are relatively flat (if the functions were exactly linear, all load distributions would be equally expensive).

Storage of Optimization Results

The evaluation of an ON/OFF scenario involves performing a local, computation-intensive load distribution calculation (SQP) using KRAWAL or neuronal networks for each time slice. If the unit selected changes in any time slice, the optimum load distribution in the other time slices remains unchanged. For this reason, the results for the time slice are stored and reused in calculating the overall cost. This is done within the optimization routine and is referred locally to one optimization run.

The output of unit commitment consists of optimal day-by-day schedules. These schedules depend on the load histories, plant conditions and ambient conditions and are stored together with these data for subsequent analysis of the optimization procedures. The data records stored in this way are relatively few (365 per year) but comprehensive. For this reason, they are stored directly and not via neural networks. Although it is unlikely that these results can be reused for further optimization runs, they provide a good starting point for initializing optimization for the following day or for a similar day of the following year.

Computation times

Calculations are performed for power plants consisting of, let's say, "N" units. Computation of the thermal input requirement in any time slice for a given load distribution requires "N" runs of block simulation (KRAWAL). These computations are performed decentrally at separate workstations in the power plant units. Optimization of the thermal input requirement in any time slice for a given unit selection calls for determination of 3 to 6 load demands per unit. For this reason, we assume 4"N" free variables for the purposes of the further analysis. Given the simple linear and quadratic dependencies of the thermal input requirement (see simulations), it is to be expected that optimization will get by with four times as many steps as variables. This is generally a very low number, but realistic in the light of the flat optimums and the low accuracy required of the optimization. At a simulation time of 8 s and 10 power plant units, an optimization time of 32 min is obtained. Thus, fine optimization of a time slice can be performed within the previous half hour using KRAWAL.

However, selection of the unit(s) to be operated requires simulation of all time slices and testing through various ON/OFF scenarios by means of a genetic algorithm. Experience has shown that a large number of function evaluations have to be performed in optimizing unit selection. By virtue of the special choice of crossover and mutation, about three optimal load distributions per evaluation of an ON/OFF-scheme have to be computed. Use in practice has shown that the load prognoses remain constant over longer time periods than half hours. Assuming therefore, 10 relevant time slices, unit commitment of 10 power plant units will require about 10^7 runs of block simulation. Since one block simulation by a neural network takes about 10^{-4} s, the whole unit commitment solution can be found within a few hours as it is required in practice.

7.4 Windage Module (Avoidance of Increased Exhaust Steam Temperature in HP and LP Turbine Sections)

Steam turbines can be operated at very low volumetric flow rates for extended periods during start-up and shut-down processes. This mode of operation leads to blading windage for HP casings (exhaust) area of reheat steam turbines. The direction of energy flow is changed in the affected stages. Mechanical energy from the turbine shaft is converted into thermal energy which increases the temperature of the steam and of the turbine components around these stages. Unallowable high temperatures can occur, which can lead to damage of the blading [19]. Failures due to turbine windage have been reported. It seems that this phenomenon might have caused damage of flow path components in large turbines of all suppliers without being clearly identified. All types of LP turbines are subject of low volumetric steam flow rate at low load, with high backpressure if there are no measures for cooling unallowable high blade temperatures can occur [20].

Task of the Windage Module

During operation of the steam turbine, heating due to windage must be maintained within allowable limits by the operating mode. Windage modules for HP and LP turbines would provide the operator with an estimation of the blade temperature at the respective turbine stages and show the expansion lines on the monitor. Its interactive user interface displays the real-time value and a trend graph of these values. Its primary task is to provide an estimation of the blade temperature, and a supervisory recommendation may be deducted from these temperatures.

In new large Siemens/KWU steam turbines temperature measurements are installed at the respective stages of the HP and LP casings. This measurement would provide an indicator to the operator whenever a certain blade temperature exceeds its limit. In addition, a neural network-algorithmic scheme for estimating the blade temperature in adjacent stages based on these measurements, could be an advantage.

For steam turbines with no “direct blade measurement devices”, a neural network-algorithmic scheme for estimating the blade temperature in particular stages based on other measurements, which are commonly available, is proven advantageous.

7.4.1 Thermodynamic Variables and Operating Data Influencing Blading Windage

The following influencing variables can give rise to blading windage:

- Volumetric Flow Rate
  Blading windage occurs at an insufficient volumetric flow ratio (volumetric flow rate/design volumetric flow rate) through an expansion section of the steam turbine consisting of several blading stages. The lower the volumetric

flow rate through a blading stage is, the worse is the inner efficiency. At a very low volumetric flow rate, the steam flows through the stage without being directed and must be accelerated to the circumferential velocity of the moving blades, which results in a braking effect. With decreasing volumetric flow rate, the output reaches zero and then changes to power uptake due to blading windage. Blading windage begins in an expansion section in the last stage and propagates back through other stages of blading with decreasing flow rate.

- Steam Density, Pressure and Dimensions
  Windage heating, which occurs at a low volumetric flow rate in an expansion section of the steam turbine, is a function of the windage power (P):

$$P = \mathrm{C} \cdot D_{\mathrm{m}}^{4} \cdot h \cdot n^{3} \cdot p$$

D_{m} Mean Diameter of Stage
h Blade Height
n Speed
p Steam Density
C Constant, depends on the geometric design of the blading

Due to the proportional relationship between steam density and steam pressure, the steam pressures in expansion sections where blading windage occurs can be used to determine the order of magnitude of windage heating. The windage must be determined for each stage and stage windage must be added to calculate the total turbine section windage produced. In multi-stage turbines, windage begins in the last blade row and extends with decreasing specific flow to the first blade rows. The last stage of a LP turbine has the highest windage losses due to the large blade length in comparison to the other stages of the LP section.

HP Turbine

Blading windage can occur in the rear area of the expansion section of the HP turbine if the steam pressure in or downstream of the turbine is high. Blading windage can occur in the following operating modes:

▷ No-load operation or operation at low load during turbine generator start-up,

▷ Low-load operation or operation at auxiliary power output following load rejection,

▷ The danger of unallowable windage heating is particularly pronounced following load rejection, because at the same time the pressure in the reheater can be temporarily high and turbine generator speed may be temporarily higher than rated speed,

▷ Testing the turbine at overspeed,

▷ Faulty opening of main steam, reheat or bypass valves.

Preventive Measures

In larger steam turbines the unallowable heating is determined by measuring the steam temperature in a representative stage. Smaller steam turbines particularly of older designs do not have these measurements. They have only a temperature measuring device at the exhaust area of the turbine casing. Other measures for avoiding overheating are:

▷ Increasing turbine output,

▷ Installation of an automatic trimming device to increase the steam flow rate through the HP turbine,

▷ Pressure releasing of the HP turbine by opening the drain valves or equipping the turbine with a separate line to the condenser with a pressure relief valve,

▷ Cooling and pressure reduction by opening the reverse flow valve for plants without bypass system,

▷ Pressure reduction with the reheater pressure relief valve for plants without bypass systems.

IP Turbine

In heating steam turbine IP casings with butterfly valves in the cross-over pipe, blading windage with unallowable heating of the blading in the lower load range can occur in heating operation with throttle condition of these valves. For measuring unallowable blade temperature, sensors in the cross-over line or in the exhaust steam nozzles can be installed. Preventive measures are:

▷ Increasing turbine output,

▷ Reducing steam pressure downstream of the IP turbine by opening the valves in the cross-over pipe.

LP Turbine

Blading windage, which can cause unallowable heating of the blading and the casing, can also be caused in LP turbines by insufficient volumetric flow rate through the blading in specific operating modes, in some cases in connection with high condenser pressure. Particularly in turbine generators with heat extraction, the volumetric flow rate through the LP turbine(s) is reduced by throttling the butterfly valves in the cross-over pipes to such an extent that high temperatures by blading windage occur during steady-state power operation.

Preventive Measures

For LP turbines with large last stage blading unallowable heating of the outer casing is measured by temperature sensors in the exhaust hood. Heating of the last stage blading is measured in the area of the last blade row. Older plants do not have sensors for measuring blade temperatures. Other measures for avoiding overheating are:

- ▷ Increased turbine output,
- ▷ Limitation of maximum allowable condenser pressure in motor operation particularly for single shaft designs of gas and steam turbine generator sets in combined-cycle plants. The steam turbine is brought up to rated speed by the gas turbine on start-up and is operated in windage mode without through flow until the heat recovery steam generator supplies steam,
- ▷ Cooling by condensate injection in the LP exhaust hood,
- ▷ Extraction cooling (special scope of supply),
- ▷ Injection of a steam water mixture into the extraction line.

Turbine Trip

Blading windage causes temperatures to rise at specific casing temperature measurement positions. The technical specification of the steam turbine manufacturer gives limits for these temperatures. If these limits are reached at certain unfavorable conditions, turbine trip must be initiated by manual or protection command.

7.4.2 Concept of the Windage Module

A lack of an effective way for monitoring the blade temperature has led into the development of a hybrid "Neural Network"-algorithmic blade temperature estimator. This tool utilizes the commonly available measurement values. The training data includes both data generated by mathematical model and experiment. The module consists of levels as shown in Figure 7.4.2.0. The administration level handles the data request from the windage process control by propagating the request at the right format to the communication level. It also manages the incoming data in certain format and forwards the data back to the process control for storage.

The action level controls the continuous background process and computation. These include the initiation of data request, data storage, all computation processes, and storage of results. This level may also include the output management which tests the validity of the computation result. The results of the ANN modules are always compared to the result of the analytical module. This verification is required to detect possible bad results which are usually caused by input values which are far away from all samples that had been presented during the training period. Large discrepancies may indicate that further retraining is in order.

The presentation level provides a graphical user interface which allows the user to view all the necessary information in several different fashions, e.g. current values and trend values. The information is presented in several layers starting with the main windage screen which will only show the blade temperatures. The subsequent layers will show the detail conditions for each

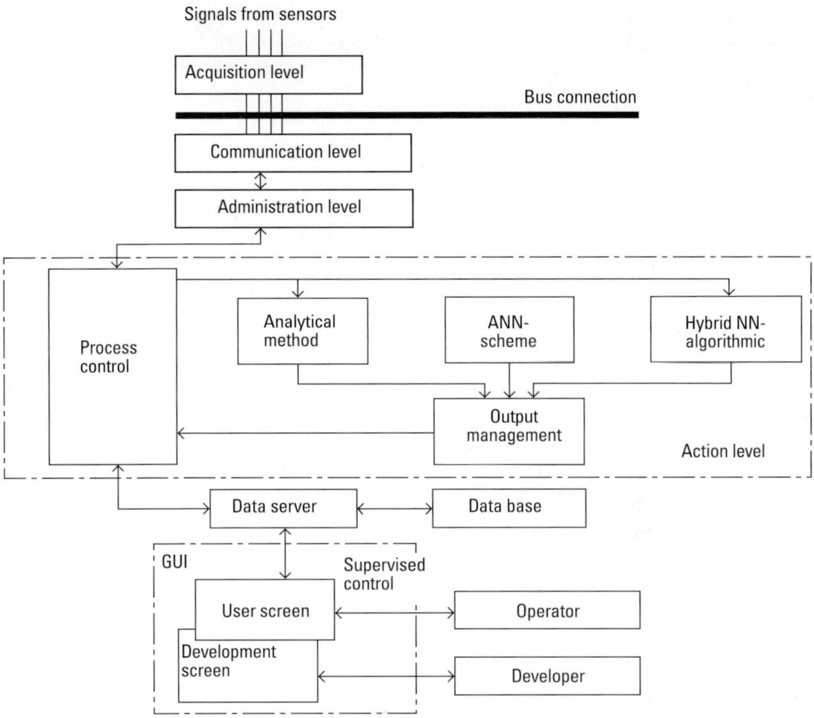

Figure 7.4.2.0 General Structure of the Windage Module

turbine section. They will provide information on all parameter values which are relevant for the operator for making the decision concerning the turbine operation.

The main structure, shown in Figure 7.4.2.0, is designed to provide monitoring functions for the user. This monitoring scheme requires system network parameters which are obtained by training the network using the appropriate data. An off-line training scheme will produce the necessary weight and other network parameters. Access to the internal module's parameters and training process may be allowed by providing the development screen. The monitoring process may not be necessary to cycle at the same rate for all turbine operating conditions. Several scenarios can be designed for specific turbines. An example of a possible operating scenario is as follows:

- ▷ Monitoring during no and low loads, or slow change to achieve no-load operation,
- ▷ Monitoring under no-load condition following a start-up,
- ▷ Monitoring during no-load operation following a load rejection.

The monitoring cycle is adjusted automatically for different conditions, and the respective display may pop up to assist the operator.

Blade Temperature Estimation

In the hybrid "Artificial Neural Network" (ANN)-algorithmic module (Figure 7.4.2.1) the blade temperature is calculated. The network is trained using the simulated data obtained by computing the estimated temperature using the analytical means for the expected normal operating domain, and actual data obtained from field experiments. This arrangement is expected to be able to estimate the blade temperature for the entire turbine operating ranges. Minimal inputs to the estimator are the real-time measurement values of the pressure of the main steam (P_{ms}), temperature of the main steam (T_{ms}), pressure of the third stage (P_{3rd}) and exhaust pressure (P_{ex}). Other possible inputs are also being considered and evaluated.

Graphical User Interface

In addition to the current values and trend values, the user interface will also show the turbine conditions within the steam behavior diagram. In the HP turbine, as there is no steam flow through the turbine following a trip, the extent of energy transfer depends on the pressure and the steam density in the turbine. At a full load trip, the corresponding high cold reheater pressure will initially be present. In order to avoid impermissible heating by windage losses, an adequate pressure decay or a certain cooling steam flow is required.

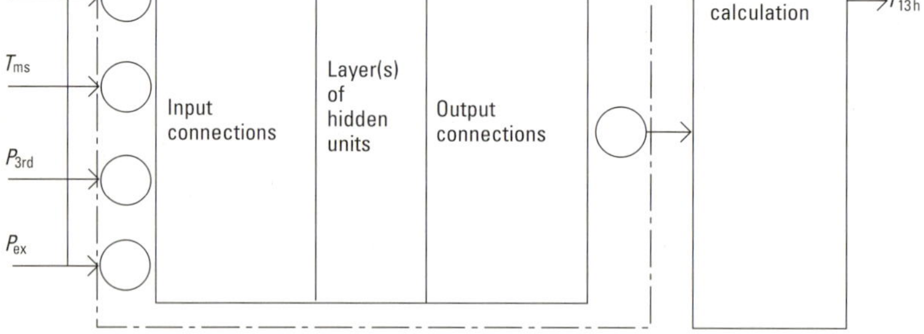

Figure 7.4.2.1 Input/Output Structure of the Hybrid ANN Module

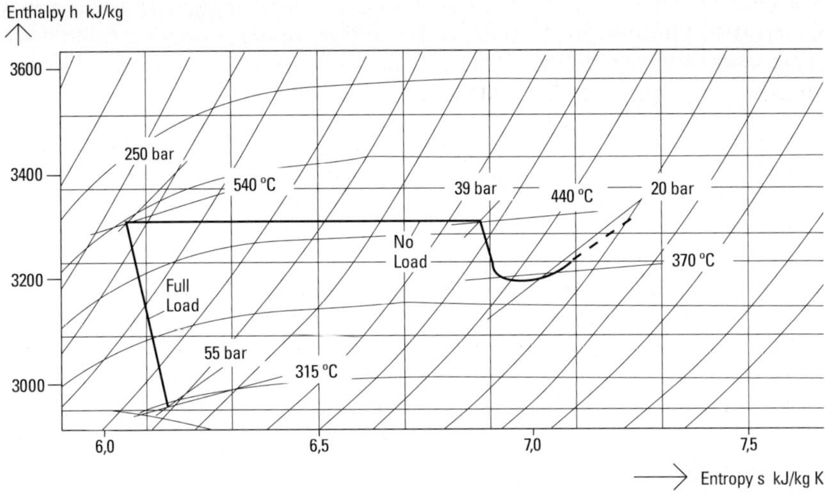

Figure 7.4.2.2 HP Turbine Expansion Lines for Full and No Load

The expansion lines in the Mollier diagram shown in Figure 7.4.2.2 for constant pressure at full-load and no-load operation show the advantage of a sufficient HP turbine flow after full load rejection to zero load. The operator will be informed by such a Figure.

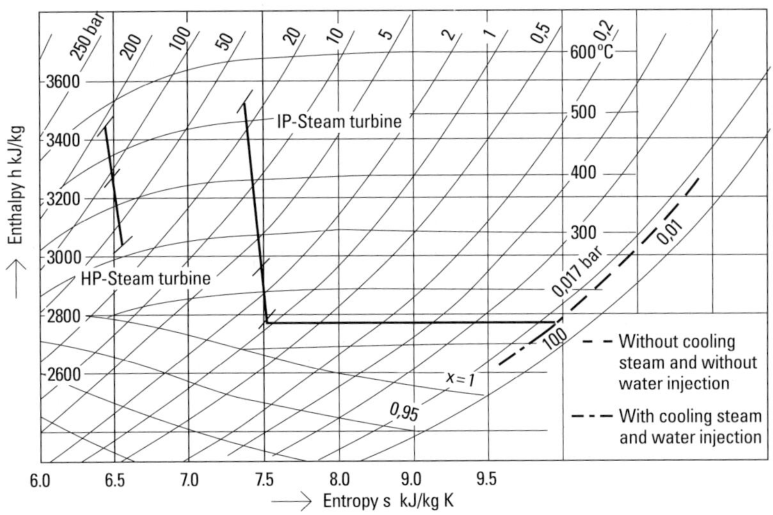

Figure 7.4.2.3
LP Turbine Expansion Lines with closed Control Valves in the Cross-Over Pipe

For heating steam turbines when the butterfly valves in the cross-over pipes of the IP/LP turbine are closed, the LP turbine requires cooling to hold the temperature rise caused by windage in the last stage within admissible limits. In this operating mode the steam in the LP turbine absorbs energy resulting from the windage losses which is predominantly significant within the last stages. Figure 7.4.2.3 shows the steam expansion curves of the LP turbine and the blade temperature with and without cooling. Water injection not only cools the exhaust hood, but also cools the last stages of the LP turbine.

7.5 Blade Vibration Information Module

Free standing rotating blades in steam LP turbines and gas turbines are designed at Siemens/KWU with regard to their aerodynamic and mechanical characteristics with the aid of advanced computer programs. Siemens/KWU has extensive experience in designing free standing blades and the use of inprocess model tests are valuable assets for establishing the requisite outline conditions for these designs. In spite of this, the installation of a blade vibration information system has been developed to cover any possible cases of increased stress, e.g. caused by alternation of the operating mode under worse operating conditions particularly for new developed large last stage blades.

Task of the Module

The overall objective of the non-contact blade vibration module is the timely detection of increased dynamic stressing of the last-stage blade so as to change the operating conditions which caused the increased stress before this leads to any reduction in service life. In the event that the blades are subject to life-reducing stresses before a hazardous condition can be avoided, the remaining service life of each blade is monitored continuously and an alarm is then issued for each blade at an appropriate safety margin before consumption of is fatigue life. The system represents a breakthrough in the prevention of blade damage. It constitutes a precondition for the use of enhanced-efficiency blade profiles and simultaneously increases the availability of steam turbines [21].

As the characteristics of the operating condition, which can cause increased stresses for last-stage blading, are unknown – this occurring, for example, for a short period of time but with intense magnitude – data from all last-stage blades must be continuously collected and checked. These data are then discarded in the event of normal operational stresses. For above normal stresses, the blade loading data are registered along with characteristic turbine operating data to allow the cause for the increased loadings to be determined at a later date. For stresses even greater than those just described, but which are not blade life-reducing, data is registered as previously described and a warning is simultaneously issued to the operating personnel. These warning signals that show the trend towards a hazardous operating condition are being counteracted by intervention in the operating mode (e.g. changing of condenser pressure or output).

The overall objective of the non-contact blade vibration information system is the timely detection of increased stress on last-stage blading so as to exit the operating condition which caused the increased stress before this leads to any reduction in service life. In the event that the blades are subject to life-reducing stresses before a hazardous condition can be avoided, the remaining service life of each blade is determined continuously and an alarm is then issued for each blade at an appropriate safety margin before consumption of the total fatigue life.

7.5.1 Functional Description

A procedure which has been applied for many years is that of alternating-stress measurement on one or more rotating blades using strain gauges. This concept can not be used during operation of the steam turbine generator because only a few blades can be measured and the service life for this type of equipment is very limited.

Therefore a measuring method has been developed for examining vibration behavior of all the free-standing blades of a gas or steam turbine blade wheel which does not require application of any equipment of the rotating blades. With this concept the vibrational deflection of the blade tips is measured with sensors located in the casing above the rotating blades and by means of a 100 MHz clock counter that determines the time required by each blade to cover the distance between the two sensors. If the time difference of a non-vibrating blade is subtracted from the time difference of a vibrating blade, the vibration displacement of the blade tip in the circumferential direction is obtained, provided that speed, sensor spacing and the distance between blade tips are taken into account. If blade vibration is not synchronous with speed, or a multiple thereof, the maximum positive or negative amplitude is obtained after several revolutions. Amplitude is likewise measured when a resonance point is passed through during coasting-up or down the turbine. Allocation of individual signals to each blade is accomplished by the phase sensor, which sends a pulse signal from the shaft.

If the time signal of a blade is analyzed in relation to the frequency, the alias frequencies for the true natural vibrations of the blades are obtained, as the vibration signal is only scanned one time for each revolution. If the approximate position of the natural frequencies in relation to the harmonic frequencies for running speed is known, the natural frequencies for each individual blade can be precisely determined from the frequency analyses. By definition, only the vibration modes which lead to a noticeable blade tip amplitude can provide measurable signals. In general, these are the first-to-third or -fourth natural modes of a free-standing blade.

Allocation of maximum alternating stress of the blade to deflection amplitude of the blade tip is achieved by calculating the characteristic modes and the

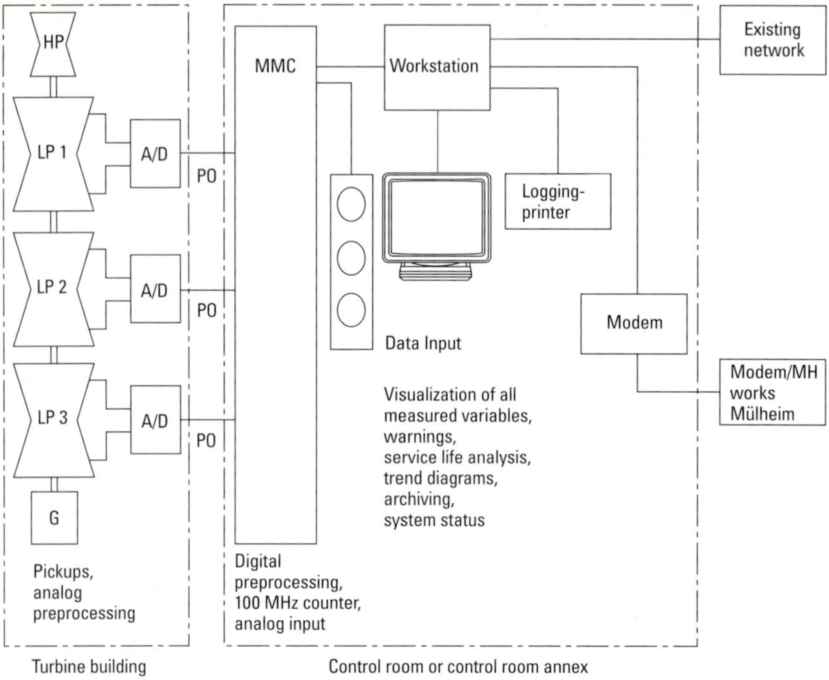

Figure 7.5.1.0 Conceptual System Arrangement of the Signal Conditioning Unit

associated alternating stress distributions. Since 1991, a blade monitoring module has been in use for blade vibration monitoring on the last stage of the LP turbines. On the basis of this module, an in-field concept has been developed for continuous vibration measurement of all the blades up to six blade wheels with up to 100 blades each. Acquisition of vibration data is performed in the following processing stages.

Signal Condition Unit

A start and stop sensor and a phase sensor are required for each row of blading for which blade vibrations are to be measured. Each electronic signal conditioner processes the vibration signals for a double row LP turbine. Digital pulse signals are generated from the amplified analog vibration signals by means of triggering. Figure 7.5.1.0 gives an overview of the system. A phase signal is also processed in order to calculate the absolute shaft position. A signal monitoring circuit is integrated in the electronic signal conditioner. This circuit checks the incoming signals and detects errors caused by missing signals or disturbance pulses. The analog signal conditioning units are tested and balanced during manufacture. The settings are optimized for power plant operating requirements. After conditioning the vibration signals are then

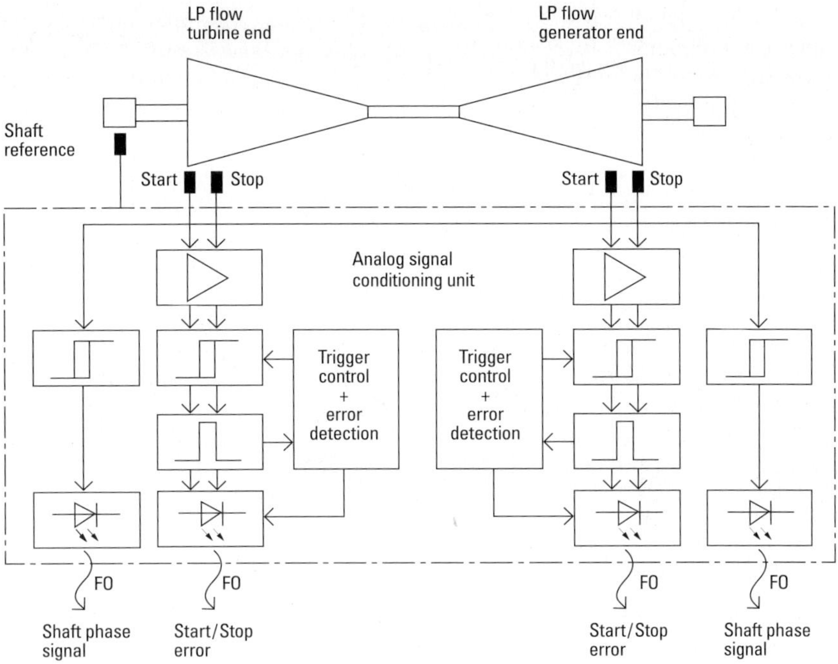

Figure 7.5.1.1 Block Diagram of the Analog Signal Conditioning Unit

transmitted in the form of digital pulses across twin fiber optic cables (one for each LP turbine flow) to the digital signal processing unit. The maximum transmission distance is 1,000 meters.

Digital Signal Pre-Processing

The signals for a LP turbine that are provided by the analog processing unit are linked to a signal processing unit in the computer cabinet (Figure 7.5.1.1). This signal processing unit comprises a powerful signal processor for each LP turbine which processes the start and stop signals in real time. With the aid of a 100 MHz counter, the start and stop signals from the 2 separate LP turbine flows are converted to time differentials whose resolution is 10 ns (10^{-6}s) and then converted to vibration displacement amplitudes for the blade tips using the circumferential velocity. An integrated monitoring logic circuit ensures that when erroneous signals are received from the turbine, the entire measurement is marked invalid and is replaced by the last previous correct series of measurements.

The data records of up to 100 revolutions of both turbine blade rows are stored in the main memory of the signal processor. Three trigger criteria compare these data records with configurable vibration limits. When one of these criteria is violated, transfer of the data record to the downstream analysis and memory computer is triggered. If one or more optional analog modules are inserted, up to 16 analog channels can be read in step with shaft rotation by each module. The number of scans for each revolution can be selected between 1 and 32.

Data Transmission Medium and Presentation

A fiber optics modem is connected to the IEC bus of the digital signal preprocessing unit. The parallel data are converted bidirectionally via single-mode fibers. All data of up to 3 parallel signal pre-processing units are transferred to the workstation via this fiber optic interface. The user can choose from various presentation formats for an instant overview of the current blade condition. In addition to current blade stressing, the system automatically calculates the service life consumption (Miner Sum) for each separate blade if the endurance limit is exceeded. The system is also able to display measured data in a trend plot for hours, days or even months. The evaluated results from the measured data can be used during the next major inspection and overhaul for a targeted search for signs of fatigue and wear.

Analysis and Storage Computer

Vibration data is transferred via the fiber-optic link by the signal processing unit only when triggered. During normal operation, no data is transmitted through the fiber optics link to the analysis and storage computer. However, as data transmission is designed for continuous storage, continual recording of measured vibration data is also possible for a trigger threshold of zero. The data at the bidirectional fiber optics interface is reconverted to IEC bus specification in the analysis and storage computer by means of a second fiber optic modem. An advanced high-power workstation with adequate memory has been selected for use as the analysis and storage computer in connection with other data of the plant condition. Data stored after triggering can be transferred by modem to our laboratories for Engineering, Testing and Mechanics for in-depth analysis.

System Triggering and Signaling

Three trigger criteria with two trigger thresholds are applicable for implementation.

Threshold 1

Blade vibration signals exceed the amplitude which corresponds to approximately 33% of the fatigue limit.

Result: The vibration signals for all blades in all the blade wheels are stored in the memory for approximately 1 minute, including 20 seconds pre-event history, together with the characteristic analog variables. An appropriate message is issued to the control room personnel.

Threshold 2

Approximately 75% of the fatigue limit has been exceeded.

Result: Same as for threshold 1. A warning is issued to control room personnel signaling that the cause of increased blade vibration must be eliminated, for example, by changing condenser pressure or output. Analysis of vibration signals and associated analog variables should be implemented without delay.

Criterion 1: Thresholds are applied for each individual blade of all the blade rows.

Criterion 2: Thresholds are applied to ½ the difference between the maximum positive and negative amplitude of two blades of a blade row for one revolution so as to detect system vibrations which are synchronous with speed. This criterion must be applied over one complete revolution.

Criterion 3: Blade vibration signals are recorded analog with the analog variables for approximately 1 minute for example every 10 minutes, regardless of the momentary magnitude of amplitude.

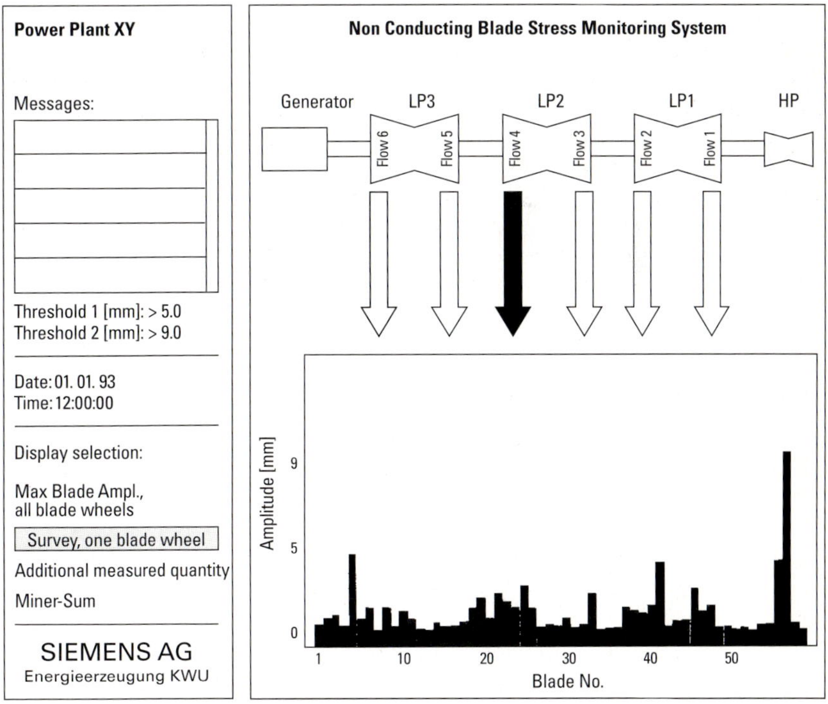

Figure 7.5.1.2 Amplitudes of Individual Blades

Using a menu, the maximum amplitudes of all the blades of one stage can be displayed with a time window of approximately 1 minute during one or more blades exceeded a trigger threshold. As an example, Figure 7.5.1.2 (derived from a hardcopy) shows the amplitudes of the individual blades. In the event that the fatigue limit of one or more blades is exceeded during a suddenly occurring event, the relative consumption of service life is calculated for each blade from the stress magnitude and the duration of stress (Miner Sum). On consumption of 75% of the service life of a blade, a warning is issued to the control room personnel. A trend display can also be generated at any time from the 10 minute values of the associated statistical and dynamic analog variables.

Miner Sum Calculation

The Miner Sum provides a powerful means of identifying the condition of a material. The Miner Sum is used to make an immediate statement on the service life consumption of the blade material. The values of the Miner Sum are in the range from 0 to 1. A newly manufactured material has a value of 0 and a fatigued material has a value of 1. The Miner Sum is only calculated when exceeding the endurance limit of the blade material. Vibrations below the endurance limit have no effect on the expected service life of the blade.

The Wöhler diagram (Figure 7.5.1.3) provides the basis for calculating the Miner Sum. Above an endurance limit each stress event (blade deflection) is assigned to a number denoting the maximum permissible number of stress cycles. Since a continuous value of the stressing (blade deflection) to a permissible stress cycle number is not possible, classes Ai are generated. A permissible stress cycle number N_{tol} is assigned to each class. Figure 7.5.1.3 shows the stress classification. The amplitude $a = \frac{1}{2} (Y_{max} - Y_{min})$ measured over the period Δt is assigned to a class Ai. Previously conducted experiments have shown that for a period of $\Delta t = 1$ s, the amplitude error due to inadequate sampling is less than 1%. It is possible to calculate the service life-reducing stress cycle number N_{actual} by multiplying Δt by the blade frequency.

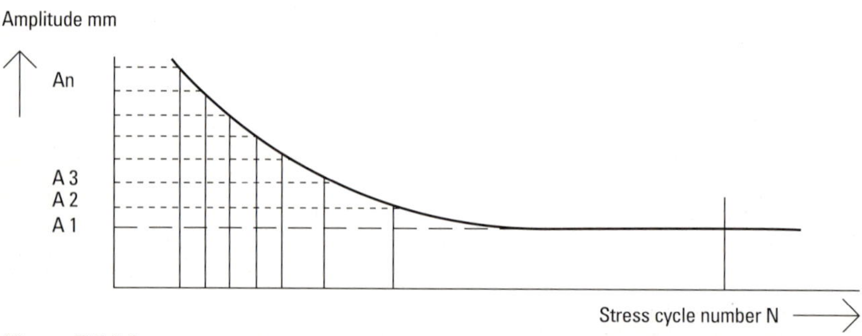

Figure 7.5.1.3
Wöhler Diagram Converted to Blade Deflection in the Circumferntial Direction

Hardware Concept

The signal processing is performed on a Siemens computer, model MMC 216. The user interface and the results of the analysis are incorporated in the workstation (see Figure 7.5.1.0).

7.6 Condenser Analysis Module

In recent years, high-integrity condensers have been developed in order to avoid corrosion-inducing influences on the water/steam cycle components as a result of circulating water in-leakage. In the past, however, the mode of condenser operation with regard to economic considerations has not always been given the necessary priority. Even today, the control of microbiological growth in the tubing of the condensers and macrofouling by marine life has still not been satisfactorily implemented in a lot of power plants, especially in those with sea water cooling. The resultant loss of plant output can be significant. A further factor to be taken into account in assessing whether condensers are being operated at their economic optimum is the effect of changes in circulating water temperature as a result of climatic conditions. Precise determination of the reference condenser pressure under all operating conditions thus requires accurate recording and conversion of the influencing parameters involved [22]. Incorporation of the auxiliary systems into the monitoring concept is also indispensable with a view to recognize incipient faults and analyzing disturbance factors.

Task of the Module

The aim of the module is to evaluate the thermodynamic behavior of the condenser under actual operating conditions. A comparison of the actual values with their respective settings allows an assessment to be made in terms of condenser pressure, turbine generator output and numerous other influencing factors. The auxiliary systems are part of the monitoring concept and the analyses performed are included.

7.6.1 Thermodynamics

The characteristic parameters of the systems are monitored in order to ensure that the condensers perform their functions reliably and thermodynamically optimal and to enable faults in the condensers and in the auxiliary systems to be detected. If the operating values of the system parameters exceed a certain tolerance range, analyses will be performed. The heat flux carried over in the condenser can be determined either by means of a heat balance or by conversion on the basis of electrical output at the generator terminals. The method on the basis of the electrical output is not feasible for heating steam turbine units or for turbine generator sets from which process steam is extracted. In such plants, it is practicable to determine the circulating water mass flow, e.g. through the circulating water pump characteristics or other measures.

The heat flux carried over in the condenser can be calculated from the heat balance of the input and the output of the heat flux and the power output. If the circulating water mass flow diverges from its setting, a change occurs in the circulating water velocity. A variation in the circulating water velocity produces changes in the pressure losses on the circulating water side. Differential pressure sensors are installed in each half of the condenser in order to account for the influences from the circulating water side.

Pressure-measuring instruments with high accuracy are not viable for measuring the condenser pressure due to their inconvenience in long-term operation. The condenser pressure is determined via measuring the saturated steam temperature as measured by special temperature measuring instruments in certain areas of the condenser. The pressures in each of the condensers are determined by converting the readings from the saturated steam temperature measuring instruments on the basis of the steam table.

Only in the lower part-load range, particularly for heating steam turbines in which localized superheating of the exhaust steam may occur, pressure-measuring instruments are used. In this case, it is not essential to plant operating economy to be able to determine the condenser pressure exactly.

The following parameters must be known for evaluation of the condenser:

Terminal output (P_{GO})	Circulating water inlet temperature (ϑ_{CWI})
Condenser pressure (p_C)	Circulating water outlet temperature (ϑ_{CWO})

The circulating water inlet temperature is measured immediately upstream of the condenser inlet. The circulating water temperature influences the condenser pressure directly. The temperature of the circulating water leaving the condenser should be measured as far downstream of the water box outlet as possible, but before any secondary cooling waters are added. Due to the temperature distribution of the circulating water, 3 to 6 temperature measuring instruments should be provided for each train, depending on the size of the train. Analysis of the condenser parameters provides an overview of the functional correlations of the thermodynamic characteristics of the condensers. The correlation between output and condenser pressure is shown below:

Condenser Pressure

Measured: $p_{C_m} = p'' = f(\vartheta_{C_m})$

Expected: $p_{C_e} = p'' = f(\vartheta_{C_e})$

$$\vartheta_{C_e} = \frac{e^{\alpha} \cdot \vartheta_{CWO_m} - \vartheta_{CWI_m}}{e^{\alpha} - 1}$$

$$\alpha = \frac{A_C \cdot K_{C_e}}{\dot{m}_{CW_m} \cdot c_{pCW_m}}$$

Difference in generator output:

$$\Delta P_{GO} = f\,(p_{C_m} - p_{C_e})$$

Heat Transfer Coefficient

The heat transfer coefficient is an important parameter for evaluating the quality of a condenser. A comparison is made between the specified heat transfer coefficient and the actual heat transfer coefficient as determined either in accordance with HEI or, for example, on the basis of our own design principles. The heat transfer coefficient does not have any direct effect on the output of the turbine generator unit, but influences it indirectly via its effect on the condenser pressure.

Heat transfer coefficient

Measured: $$K_{C_m} = \frac{\dot{Q}_{C_m}}{A_C \cdot LMTD_m}$$

$$LMTD_m = \frac{\vartheta_{CWO_m} - \vartheta_{CWI_m}}{log\,e\,\dfrac{\vartheta_{C_m} - \vartheta_{CWI_m}}{\vartheta_{C_m} - \vartheta_{CWO_m}}}$$

Expected: $K_{C_e} = f\,(\vartheta_{CWI_m} \text{ and } c_{CW_m})$

Deviation: $\Delta K_C = K_{C_m} - K_{C_e}$

Terminal Temperature Difference

The terminal temperature difference is a measure of the quality of a condenser. The terminal temperature difference determined by the program is compared with that specified, discrepancies are monitored, and an analysis is performed to determine the cause of any change.

Measured: $\Delta\vartheta_{TTD_m} = \vartheta_{C_m} - \vartheta_{CWO_m}$

Expected: $\Delta\vartheta_{TTD_e} = \vartheta_{C_e} - \vartheta_{CWO_m}$

Deviation: $\Delta\Delta\vartheta_{TTD} = \Delta\vartheta_{TTD_m} - \Delta\vartheta_{TTD_e}$

Condensate Temperature in the Hotwell

Measurement of condensate temperature must take into account that various drains may enter the hotwell at different temperatures. For this reason, the instrumentation points should be carefully located so as to give a representative reading of the condensate temperature. Subcooling of the condensate produces heat losses. The additional heat input required to heat up the subcooled condensate in the feed heater results in a loss of turbine generator unit output. Condensate subcooling can also occur at low cooling water temperatures if the steam turbine goes into choking condition. Inadmissible deviations will be annunciated to avoid power losses.

Condensate Subcooling

Measured: $\Delta\vartheta_{C_m} = \vartheta_{C_m} - \vartheta_{COND_m}$

Expected: $\Delta\vartheta_{CS_e}$ (given by design)

Difference in Generator Output:

$$\Delta P_{GO} = f\,(\Delta\vartheta_{CS_m} - \Delta\vartheta_{CS_e})$$

Terms and Indices

The terms and indices used for the thermodynamics are depicted as follows:

P, P_{GO}	Terminal output
Q	Heat flow
Q_C	Condenser duty
m_{CW}	Circulation water mass flow
p_C	Condenser pressure
p''	Saturation pressure corresponding to condensing steam temperature
ϑ_{COND}	Condensate temperature in the hotwell
ϑ_{CWI}	Circulation water inlet temperature
ϑ_{CWO}	Circulating water outlet temperature
ϑ_C	Condensing steam temperature
$\Delta\vartheta_{CS}$	Condensate subcooling
$\Delta\vartheta_{TTD}$	Terminal temperature difference
LMTD	Logarithmic mean temperature difference
$c_{p\,CW}$	Circulating water specific heat capacity
c_{CW}	Circulating water velocity
A_C	Condenser surface
K_C	Heat transfer coefficient
e	2,7183
Index	m = measured, e = expected

7.6.2 Air Extraction Units

The vacuum pumps serve for hogging operation during start-up and holding operation. During holding operation the air intruding into the condenser has to be extracted, together with the water vapor bound to the air for physical reasons, even at the lowest condenser pressure. If these requirements are not fulfilled, the resulting increase in the condenser pressure reduces the turbine generator unit output. Analysis of the air extraction system is explained in the following, taking the commonly used liquid ring vacuum pump as an example.

Continuous measurement of intruding air by means of thermal anemometers leaving the ELMO® vacuum pumps enables constant monitoring of air leaks.

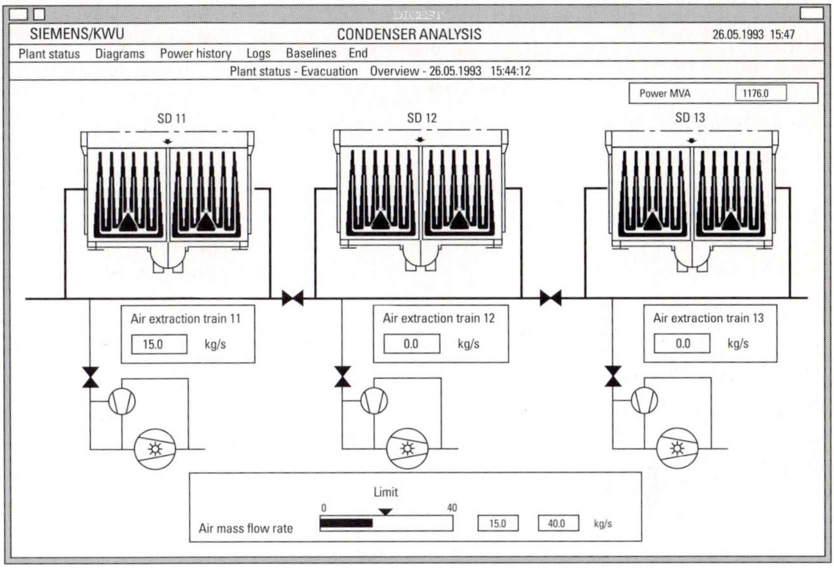

Figure 7.6.2.0 Condenser Tubing Array with the Air Extraction System

Monitoring of the measured values (Figure 7.6.2.0 derived from a hardcopy) and comparison with the defined references allows early detection of disturbances, especially with the aid of trend analyses. Continuous air leakage monitoring also makes it possible to carry on normal operation with a single pump in certain cases, reducing the station service power requirement considerably. With the aid of this module it is possible, not only to identify air ingress but also to evaluate the performance of the gas ejectors and possibly to detect cavitation in the ELMO pumps and to monitor the proper functioning of the ELMO water coolers.

7.6.3 The Condenser Analysis Sequence

The condition of a condenser must be calibrated against the calculated values, and the process parameters are subject to change in the course of prolonged operating periods. In order to account for these changing circumstances, the condenser analysis automatically adapts the reference status to the actual status. This also makes it possible to adapt to new plant conditions such as arise, of values, e.g. following an extensive overhaul of the turbine generator unit. This update is performed when a person with the appropriate access authorization changes the baseline or initiates a new baseline measurement by means of a mouse click. Following extensive plausibility checks of the individual measured values, an analysis is performed when the measured values have stabilized. The criteria for determining when this is the case, a number of measurement and process variables which must lie within specified tolerance bands.

Due to the high information density and the advantages of having a large storage capacity (> 1 year), the data must be compressed, that is to say, the scope of the data accumulating under normal conditions must be limited. This is achieved in the condenser analysis by storing measured and process data only when selected process variables leave a specified band width. Diagrams and figures are provided for the various thermodynamic parameters to give an at-a-glance overview of the status of the condenser. The auxiliary systems are also represented in displays to give the operating personnel a complete overview of their functioning and status.

Figure 7.6.3.0 shows the sequence of the condenser analysis and diagnosis initiation (see also chapter 5.3 for diagnosis matrix). The display of the deviations in the significant status parameters provides the operator with a rapid overview of the magnitude of the deviations and indicates whether corrective actions would be cost-effective. Suitable system displays are provided for this purpose. At low condenser pressures, the high steam velocity at the swallowing capacity of the final stage of the LP turbine leaves no room for further increasing output by further reducing condenser pressure (choking condition). Under these operating conditions, output can be maximized by reducing the circulating water mass flow rate, thus reducing the station service power requirement for the circulating water pumps. Displays indicating this status are shown on screen [23].

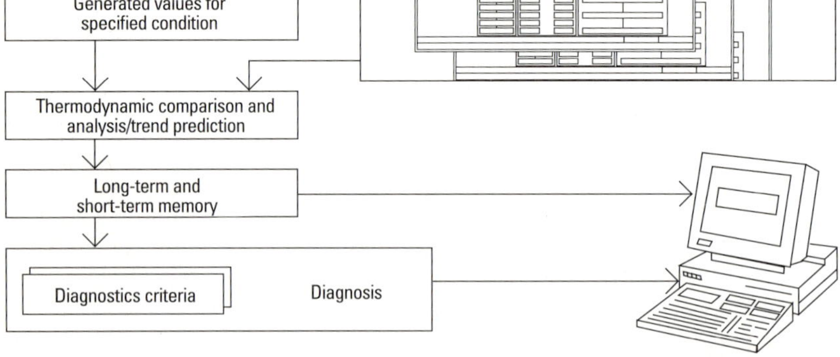

Figure 7.6.3.0 Condenser Module, Program Sequence

The integrity of the condenser is of major importance for the operation of the turbine generator unit. Circulating water in-leakages impair the quality of the feed water and may in extreme cases make it necessary to shut down the affected condenser half, with associated loss of output. In order to avoid operation in harmful ranges (generally extreme operating ranges with wide circulating water fluctuations) in which droplet impact erosion may occur, the corresponding operating conditions in which water droplet erosion can occur are calculated and the time spent operating in these dangerous ranges is monitored. This makes it possible to gain an indication of the probability of leaks as a function of the operating hours run in harmful operating ranges. Experience on which to base evaluation of the erosion hazard potential is available.

The instrumentation and control equipment required for condenser analysis consists primarily of the interface between the process and the user. Within the I&C hierarchy for power plants, these functions are implemented at the equipment group and equipment unit I&C levels and at the process control level. In a power plant that is already in service, the majority of the parameter readings required for condenser analysis are, as a rule, already available in the form of on-line measurement signals. These signals are combined, processed and input into the analysis by the equipment group and equipment unit level I&C. Any further instrumentation signals required for the specific plant, primarily from the turbine building area, are processed via signal conditioning by means of the SIMATIC S5 – in the same way as the signals already present – and provided to the workstation through the SINEC H1 bus.

7.7 Bar Temperature Analysis Module for Generators with Water-Cooled Stator Windings

As a rule, a fault in the cooling system will cause overheating and destruction of insulation, occasionally even melting of active metal parts. Special analysis procedure such as stator bar temperature monitoring for generators with water-cooled stator windings (THDF series) makes it possible, in conjunction and linked up with other influencing parameter measurements, to detect and pinpoint zones with disturbed cooling [24]. Action can be taken to prevent the damage spreading and in most cases it will even be possible to continue plant operation at a suitable, adapted output level. Figure 7.7.0 provides an overview of the concept.

Task of the Module

The bar temperature monitoring facility is a computer-assisted analysis system used for prompt detection of even small restrictions in the water-cooled hollow conductors of the stator windings. The system is basically intended for high-capacity generators (Type THDF/THFF). Broken strands can also be detected by the system. The module performs quasi-continuous monitoring: The water temperature in each bar is measured and analyzed in

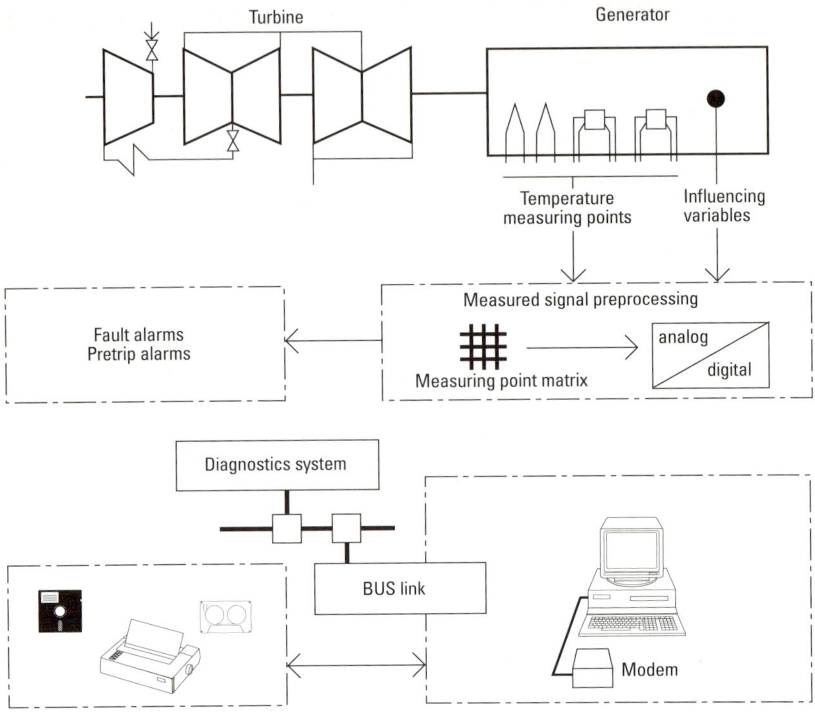

Figure 7.7.0
Generator, Individual Bar Temperature Analysis for Water-Cooled Stator Windings

half-hourly cycles, and any inadmissible deviations from the setpoint reference values are reported immediately. The measurement and analysis result can be called up at any time on the computer screen or documented in log form.

The long-term storage of changes in temperature gradients also allows long-term forecasts to be made. This module is capable to match the extent of a cooling system fault to generator output, thus allowing optimization of cooling in the perturbed case and reducing the danger of further damage. We establish the reference setpoints for bar temperature analysis using load-dependent variables. Fingerprint measurements taken on the healthy generator serve as a basis for these.

7.7.1 Structure of the System

In modern high-rating generators the stator windings which consist of individual bars are water cooled. For this reason, either all or some of the strands of a stator bar are hollow. All strands are enveloped in a common

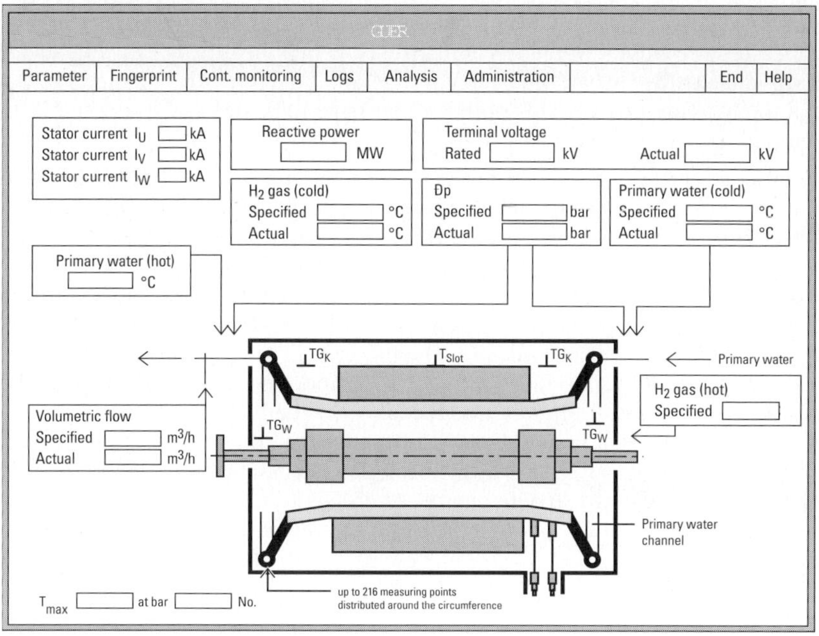

Figure 7.7.1.0 Bar Temperature Monitoring Concept

high-voltage insulation which reduces heat exchange with the environment. The cooling water flows from a cooling water header through insulating tubes to the water boxes soldered to the end of the bar. On the hot-water side, the cooling water leaves the bars through these water boxes and is conveyed to the hot-water header via insulating tubes. Standard measuring points are: Temperature of the hot primary water, temperature of the cold primary water, temperature of the hot H_2 gas, temperature of the cold H_2 gas, slot temperature, differential pressure of water across stator and volumetric flow rate of water. The cooling function is shown in Figure 7.7.1.0 (derived from a hardcopy). A winding consisting of bars with hollow and solid strands can have more than 5000 parallel flow paths in a hydraulic circuit. Only the total rise in temperature of the water (*Delta* = hot prim. water – cold prim. water) or the characteristic value of the total hydraulic resistance $R_h = Delta_p / V^n$ can be derived from the standard sensors as an indication of the changes in the windings.

A statement about individual paths, bars or even hollow strands cannot be made with these values. Hollow strand blockages not only result in changes to the rise in temperature of the water in the bars but also in hot spots which cannot be detected with measuring instruments. Figure 7.7.1.0 gives an overview of a display with the respective data ready to be called. If it were possible to measure the cooling path of each individual bar, it would be possible to detect blockages in individual hollow strands from its characteristic values. Suitable volumetric flow rate measuring instruments for individual paths in bars are not yet available. This is why changes in the volumetric flow rate of the cooling water have to be derived from the rise in temperature of the water.

In order to perform these measurements, temperature sensors have to be fitted on the hot-water side of each individual bar in addition to the standard measuring points. The fitting of such sensors has been carried out for many years. However, until now these measurements were not sufficiently evaluated because usually only a fixed limit formed the basis for monitoring the hot-water temperatures which were above the maximum expected temperature. However, with this method it is not possible to achieve satisfactory monitoring and warning of thermal endangerment because the rise in temperature of the water depends on the load and is different in every bar in the parallel circuit, due to differences in manufacturing and slot-specific magnetic field. If the flow through a bar is impeded, the volumetric flow rate is reduced and the temperature increases. However, this temperature rise is minimal in comparison to the increase in the temperature of the hot spots.

For example, a blockage of 10 % in the hollow strands of a bar will only cause a rise in the bar water temperature of 3 – 4 K, although, if the blocked hollow strands are in an unfavorable position, the permissible rise in temperature of the insulation class may already have been exceeded locally. Due to the fact that the inherent spread of the water temperature rise can exceed the permissible change of about 3.5 K and to ensure reliable monitoring also at partial load, i.e. with a lower water temperature rise, any fixed limit of water temperature rise must be abandoned in favor of a variable limit which is to be assigned specifically to each individual bar. If breaks occur in the strands, the total resistance of the bar and bar losses increase. The resulting changes in the rise in temperature can be detected to the same degree as the change in the rise in temperature resulting from blockages if appropriate evaluation methods are applied.

7.7.2 Analysis Method

Blocking of the hollow strands is analyzed by comparing the actual increase in temperature of the bar cooling water with a reference setpoint. The change in the rise in temperature is the monitoring criterion. The reference setpoint is derived from the basic measured data which have been recorded on the "healthy" generator and which are stored in a "fingerprint", taking all current and load-dependent influencing variables into consideration. A computer-assisted monitoring method which performs monitoring almost continuously was developed to solve this problem. Every individual bar water temperature is measured in half-hourly cycles, evaluated and monitored for impermissible changes. A signal is outputted if the limit is exceeded and long-term storage makes forecasts possible.

A temperature sensor (thermocouple or resistance thermometer) is located on the tube connection nipple of each bar which lies on a potential to ground and is thermally insulated against the influence of the cooling gas. The measuring lines are brought out of the generator protected against fault voltage and are connected to the central terminal unit of the measured data preprocessing equipment. Measured data are preprocessed in the SIMATIC S5 controller system. Here, up to a number of 216 bar water temperatures and all current, load-dependent process variables (influencing variables) which affect the increase in temperature of the bar water, are read in and digitized. The measured data are submitted to a plausibility check and assembled in frames.

Data transmission between the measured data preprocessing unit and the monitoring computer is performed via a SINEC H1 bus. The monitoring computer, a workstation, controls measured data acquisition and takes the digitized values from the measured data preprocessing unit. Before commissioning, a fingerprint recording is made. During this process (which involves selective measurement runs of the healthy machine) a polynomial *Delta* $T_{Ref} = K \cdot \Sigma a_i f_i$, where a_i are the coefficients and f_i the influencing variable functions is derived from the measured data for each bar.

The measurement operations are followed by the monitoring process which includes an initial four-week learning phase during normal operation. During

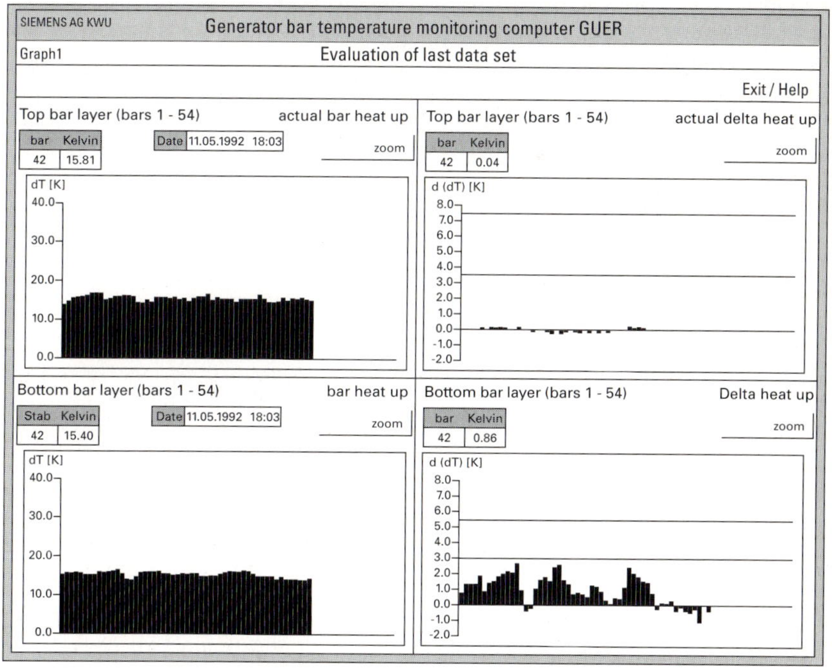

Figure 7.7.2.0 Evaluation of a Data Set, Bar Heat-Up and Delta Heat-Up

this learning phase, additional measured data to expand the fingerprint are collected. The fingerprint expanded by the learning phase can only be released by an authorized person. In monitoring mode, the computer calculates the reference setpoint for every bar water temperature rise measured and determines the current interference-related changes:

$$Delta\ (Delta\ T)_i = Delta\ T_{act} - Delta\ T_{iRef} \quad \text{(i, “Bar ID-Number”)}$$

Influencing variables are the stator current, reactive power and terminal voltage. The monitoring computer checks the change in the rise in temperature for any limit violation and also converts the change to a normalized load case. All current and normalized rises in temperature as well as the connected influencing variables are stored for one year. The stored values allow trend calculations and diagnoses. The results of this almost continuous monitoring can be called up at any time in a display or by printing out logs. The monitoring program covers:

- ▷ Reference setpoint formation for each individual bar with the help of a polynomial that is determined when the fingerprint is made,
- ▷ Continuous monitoring with setpoint actual data comparison.
  The reference temperature rise is calculated from process variables using the coefficients determined in the fingerprint and then compared with the measured rise in temperature of the bar water. The deviations are evaluated. An example is given in Figure 7.7.2.0 (derived from a hardcopy).
- ▷ Trend analyses are possible in addition to the freely configurable graphics based on the stored data (which is retained for 1 year and 1 week, 371 days) can also be used.

7.8 Monitoring (Analysis) and Diagnostics Module for Early Fault Detection on Electrical Equipment

7.8.1 Monitoring (RF Analysis) Module

Arcing caused by broken current-carrying conductors or by discharges in high-voltage equipment act as transient current surges with a broad frequency spectrum. If sustained over a long period of time, this causes irreparable damage to conductors and insulation of the high voltage (HV) components [25]. In the case of generators, for instance, the radio-frequency (RF) pulses pass through the stator winding, the distance traveled depending on the distance from the point of origin, can subsequently be measured at the generator leads via the bushings.

Dielectric measurements with respect to stator winding insulation aging can also be performed during unit operation. With the respective experience available to analyze the measuring results, it will be possible to evaluate the appropriate degree of aging as well as to determine and assess the severity of local failures. However, extensive development work of our high-voltage laboratory in the field of partial discharge measurements has produced a monitoring system for early detection of incipient faults in electrical equipment of power plant units. An overall arrangement of the module is shown in Figure 7.8.1.0.

Task of the Module

With the radio-frequency monitor Siemens/KWU is introducing a novel and exceptionally versatile monitoring system detecting and interpreting RF fault voltages. In conjunction with a dotted-line recorder, the module is suitable for

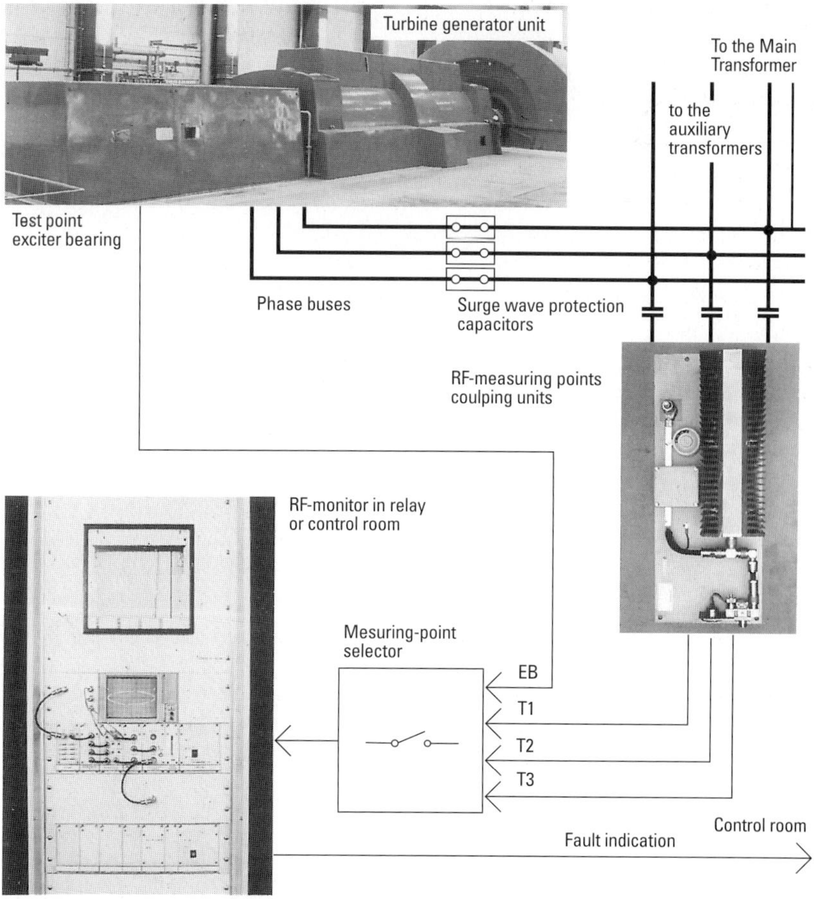

Figure 7.8.1.0
Arrangement of the RF Monitoring for High Voltage Components in Power Plants

automatic operational monitoring and recording the measured RF level. A standard oscilloscope can be connected for visual signal analysis thus enabling fault localization in the plant.

The extended version equipped with a microprocessor permits connection of any plant-specific recorder and transmission of the readings and alarm signals

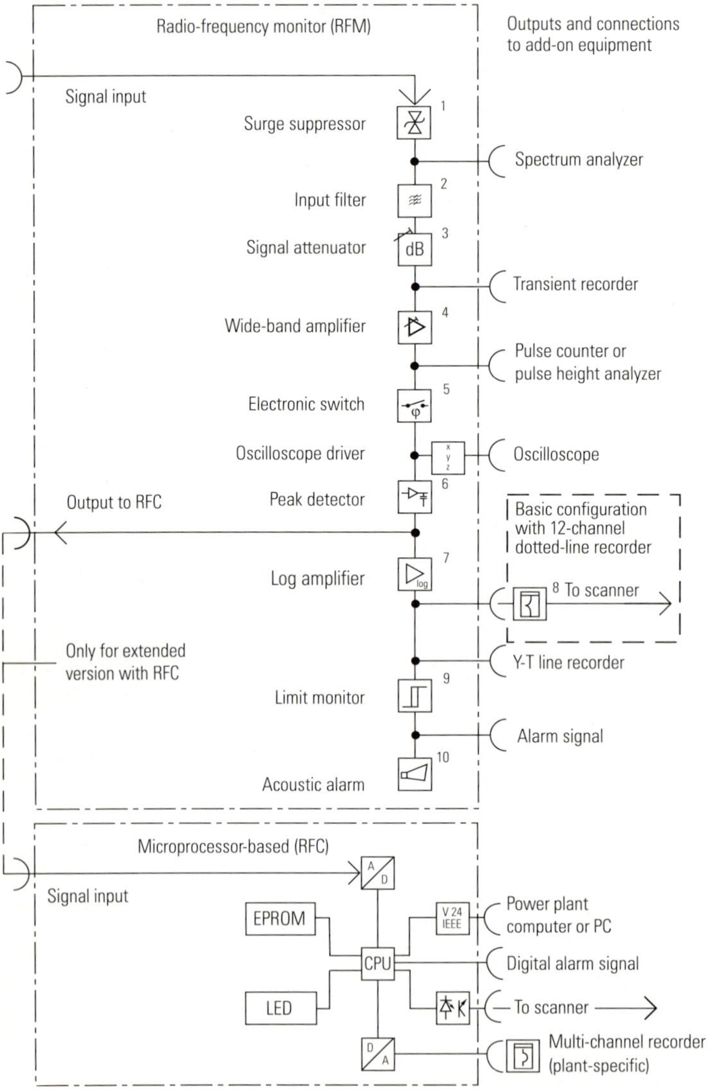

Figure 7.8.1.1 RF Monitor and Extended Version with RF Analysis

to a computer or integration of the RF monitor into a diagnostics system (described later).

Principle of Operation and Capabilities

The basic structure of the RF monitor is depicted in Figure 7.8.1.1. It consists of an input filter with a surge suppresser in the incoming circuit that limits the frequency range. An attenuator in conjunction with an amplifier permits the signal to be detected over a wide dynamic range. System-related RF signals will be suppressed by an electronic switcher, such as commutation or trigger pulses, which are present in the detected signal with a specific phase angle. For this purpose a gate circuit is employed. A peak detector with a logarithmic amplifier in outgoing circuit connected, converts the pulse amplitude into the recorder signal required for trend monitoring. A driver amplifier matches the signals to the inputs of the recorder and those of the signaling groups. The pulse amplitude is monitored visually on a LED display, with an analog alarm being initiated as soon as a present limit is exceeded. For oscilloscope display of the radio-frequency pulse signals on an elliptical time base, sinusoidal voltages derived from the system frequency are applied to the X and phase-displaced in quadrature, for the Y input. In addition, Z-brightening is synchronized with the gating.

In the basic configuration the multi-channel dotted-line recorder scans 12 instrument locations. The permanently installed scanner in the recorder activates a combination of relays which selects the instrument locations and connects them to the RF monitor for measurement.

The signals are taken at several coupling locations in the plant's power system. Preferred coupling locations are the surge capacitors in the generator phase leads (L1, L2, L3), the generator neutral (N) and the insulated exciter bearing. In this way, the generator is instrumented on several sides. Further instrument locations are on the high side of the generator transformers and the auxiliary power supply equipment can also be cyclically monitored. Instead of the 12-channel dotted-line recorder with internal scanning a microprocessor-based controller (µP-Controller, RFC) can be used which takes over scanning and hence expands the capabilities of the RF monitor considerably by:

- ▷ A/D conversion of the peak detector signals,
- ▷ Averaging over an adjustable time interval to balance out sporadic spikes,
- ▷ Logarithmic conversion,
- ▷ Selection of one of max. 12 instrument locations for continuous observation,
- ▷ Simultaneous and continuous output for all 12 instrument locations allows connection of user-specific plotters,
- ▷ Numerically evaluated digital alarm output in the event of limit violation,

▷ Serial interface (V24) for output of instrument location, reading date, time and alarm,

▷ LC-Display (bar chart) for a quick-look.

Connection to Add-On Equipment – Analysis Module

The modular design of the RF monitor also allows connection to the front outputs of a number of add-on devices for further fault recording and analysis functions. The measuring sensitivity of the monitoring system is considerably reduced by plant-specific periodic background noise due to operation such as thyristor or diode pulses. This noise is suppressed by using filters with suitable response characteristics or by means of the gate circuit which must be set on start-up of the RF monitor. Transient noise levels such as switching operations in the grid or tap-change operations of the generator transformer, are suppressed by numerical signal evaluation (averaging) in the RF controller.

Hardware

The module is installed in a rugged cabinet for use in power plants. In combination with the shielding features, this cabinet provides excellent protection against external interferences. The consistent modular design allows individual modules to be replaced and permitting continual upgrading to accommodate state-of-the-art RF-fault voltage measurement technology.

7.8.2 Diagnostics Module (RF Diagnosis)

Task of the RF Diagnostics

The RF diagnostics module is an extension of the RF monitor which has been used for years as an instrumentation and monitoring system for detecting conductor and electrical insulation faults in power plants [26]. The task fulfilled by the RF diagnostics modules is to provide support for power plant personnel in the analysis and evaluation of signals generated by the RF monitor. Figure 7.8.2.0 shows an overview of the concept. There are two possible ways to achieve the diagnosis results:

- Stage 1, off-line diagnostics
  RF diagnostics on workstation or PC permits fault diagnosis by the user in dialog mode with the resultant measurements of the RF monitor.
- Stage 2, on-line diagnostics
  Digital signal processing unit permits direct transmission of measured data characteristics into the RF diagnostics and performs on a workstation fault diagnosis fully automatically.

The central feature of the RF diagnostics is a knowledge base, in which the expert knowledge on a large number of relevant conductor and electrical insulation faults is stored. The knowledge base is fed by technology experts in the Siemens/KWU engineering generator department. The power plant

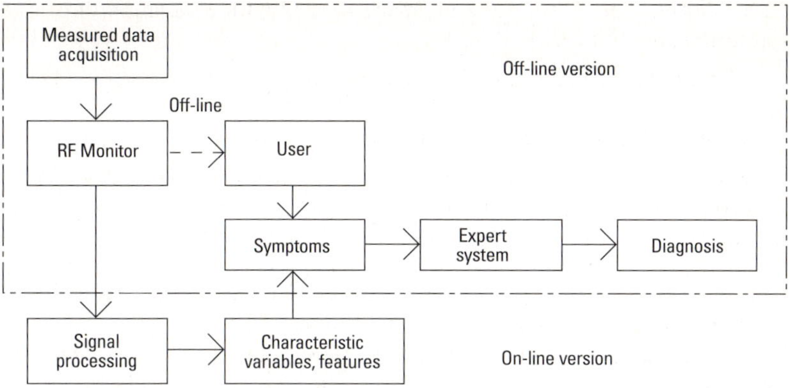

Figure 7.8.2.0 Upgrades of RF Diagnostics

personnel has access to this knowledge for purposes of fault diagnosis via user-friendly dialog. Both the generation of the knowledge base and the dialog process are designed in such a way that they do not demand any knowledge of electronic data processing on the part of the user. The RF diagnostics can be connected as an individual module to the higher-level turbine generator diagnostics system DIGEST.

Features of the Concept

With the aid of RF monitoring electrical equipment, faults in power plants are detected which occur as a result of high-voltage discharge or conductor breaks. In the case of incipient faults, the signal amplitude, the number of discharges and their phase angle are displayed as analog data. Changes in these signals, in conjunction with the comparison of signals from several measuring points in the power plant, indicate the nature and location of the cause of the fault. As an independent instrumentation and monitoring system, the RF monitor has established itself since 1985 in power plants as an expedient aid for early detection of faults and their causes. The following problems are, however, encountered in signal analysis and interpretation:

- A fault can only be detected if the RF monitor annunciates an alarm in the power plant control room or amplitude plots are observed at specific intervals, and the image displayed is detected as being a fault.
- As a result of plant-specific interferences, it may under certain circumstances be very difficult to detect the actual fault. Furthermore, images of a similar nature which can only be interpreted by an expert may have quite different causes. Particular importance is attached to the correct inter-

pretation of the pulse oscillograms of the individual measuring points and the comparison with earlier diagnoses.

- Certain faults may be indicated by the continuous change of one or several measured data over time. The problem is then the early detection of a trend.

To date, mainly experts from Siemens/KWU who have analyzed the signals together with the power plant personnel, prepared a diagnosis and given recommendations for a continued operation of the unit. This procedure has, however, proved to be very complicated. At Siemens/KWU, the boundary conditions which influence the current signals are often not adequately known, while the base of experience on other conductor and insulation faults is not adequately available in the power plant. These difficulties may result in the fact that a belated decision on further operation can only be made and that the decision may perhaps also be incorrect. The difficulties mentioned were decisive for the development of a diagnostics system which is implemented in two stages. The objectives here are:

▷ Early detection of incipient faults,

▷ Rapid analysis of causes when monitoring systems respond,

▷ Reduction of the possibilities of error in the generation of a diagnosis and

▷ Proposals for countermeasures, remedial measures and repairs.

The RF diagnostics should thus ultimately contribute to cost-effective plant operation which is expressed by increased component operating reliability and availability, and conservation of consumables and plant equipment. The problem that complex results can only be fully utilized in situ with a permanently available experience base is not restricted to the analysis of electrical insulation and conductor faults. The system presented here is a module within a complete diagnostics system for turbine generators.

Principle of the Analysis

The measured data supply the RF monitor with an amplitude display which is plotted independent of time by a multi-channel recorder. The cause of slight fluctuations in the amplitude of the same magnitude recorded by all measuring points is in most cases attributable to operational changes, e.g. active and reactive load changes. More pronounced increases in amplitude from one of more measuring points indicate in most cases a fault which is not a normal operating state, whereby the comparison of amplitude alone may already indicate the origin of the fault. In addition, a characteristic pulse image is presented in the form of a "Lissajous plot" whose cycle time is equal to the period duration of alternating current (20 ms at 50 Hz). Each of the connected measuring points supplies a "Lissajous plot" with the superposed signals. The "Lissajous plots" are subsequently analyzed. This is performed mainly by calling up experience on similar faults. Furthermore, the results of laboratory investigations are also available for comparison in establishing the fault. It is necessary to recognize and classify the pattern of the current fault in the “Lissajous plot”. The results of this work are information on the nature and location of the fault in the electrical system and a recommendation on the further procedure in the case in question.

Dialog (Diagnosis-Stage 1)

The off-line method (Figure 7.8.2.0) shows the structure of the module and the data flow. Here the user is assisted in the interpretation of the recorded amplitude and “Lissajous plot” of the RF monitor by a computer. To this end a knowledge base is stored in the computer which contains the technical knowledge of a human expert required for interpretation. The symptoms which the pulse images and recorder plot concern, e.g. amplitude, phase angle and shape of the signals, as well as the influencing variables are hierarchically structured. The relationships between the symptoms and the diagnoses are established on the basis of rules.

Since in most cases no clear-cut yes/no-statements are possible, the symptom-diagnosis link is realized by so-called confidence factors which represent a measure of the strength of this link. The knowledge acquisition component enables the expert to input symptoms and diagnoses which are linked with one another in the form of rules. The relationship thus generated is transformed with the aid of a rule generator into rules for the lower-level expert system. Following rule compilation, these rules are then available to the user in the dialog system.

The demand made by the user for a diagnosis does not take place in the form of a rigid question and answer dialog; the user himself controls this process. This enables greater flexibility and a user-friendly approach. Particular importance was therefore attached to a readily operable user interface which could be intuitively assimilated, and which essentially precludes spurious inputs and provides the user at all times with a context-dependent help system. All input options are stored in structured form as menus and need merely be selected with the mouse. Furthermore, there is also the option to call up the display for comparative representation or for validation of the diagnosis of the basis of the “Lissajous plots” stored in the computer. Figure 7.8.2.1 (derived from a hardcopy) shows an output of the dialog. The result of a diagnostics setting is the output of various diagnoses with different degrees of probability. The dialog guide component of Stage 1 essentially has the following features:

- ▷ Window technology,
- ▷ Pull-down menus,
- ▷ Context-intensive user guidance (only functions which can be activated are displayed on-screen),
- ▷ Color help texts and, as necessary, images for each function,
- ▷ Expert mode with short-form commands,
- ▷ Presentation of rule images as examples.

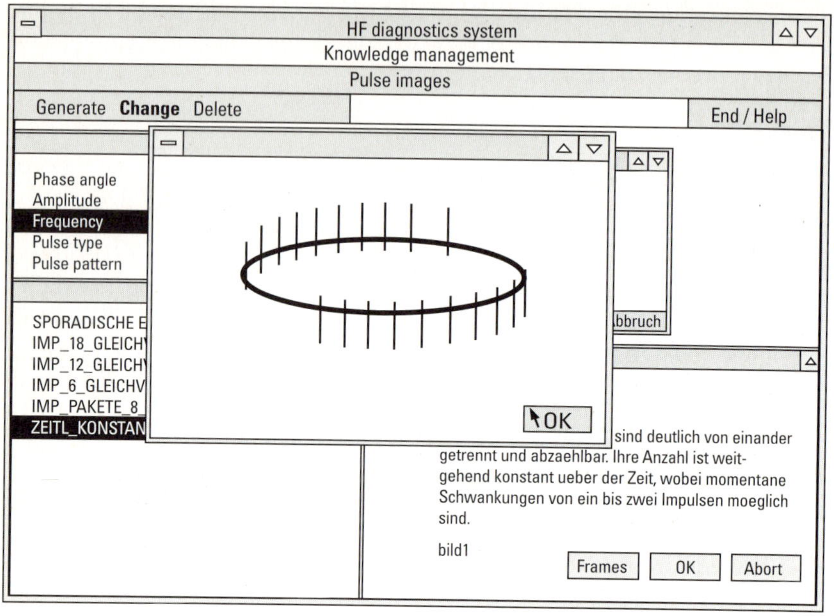

Figure 7.8.2.1 RF Diagnosis, Dialogue Guide

Historically, this concept has been developed as an appropriate extension stage to the RF monitoring in order to provide the operator with the necessary information to be able to evaluate incipient faults and disturbances in good time without consulting the manufacturer's specialists. This module can be considered as an indispensable basis for the realization of the automatic "Diagnosis Stage 2", but can also being used in conjunction with this stage in certain circumstances.

Automatic (Diagnosis Stage 2)

The second way of deriving a diagnosis from the recorded amplitude and the current "Lissajous plot" is fully automatic. Figure 7.8.2.2 shows the basic structure of the concept with the data flow included. The relevant measured data are recorded on-line. From these data the computer derives by various signal processing methods (filtering, FFT, etc.) and means of mathematical analyses characteristic variables and features which it then subsequently converts into the symptoms already described in the off-line mode.

The knowledge base mentioned earlier thus also continues to be used in the on-line system and this Stage 2 thus constitutes an extension of Stage 1. The user can continue to consult the dialog component in order, for example, to

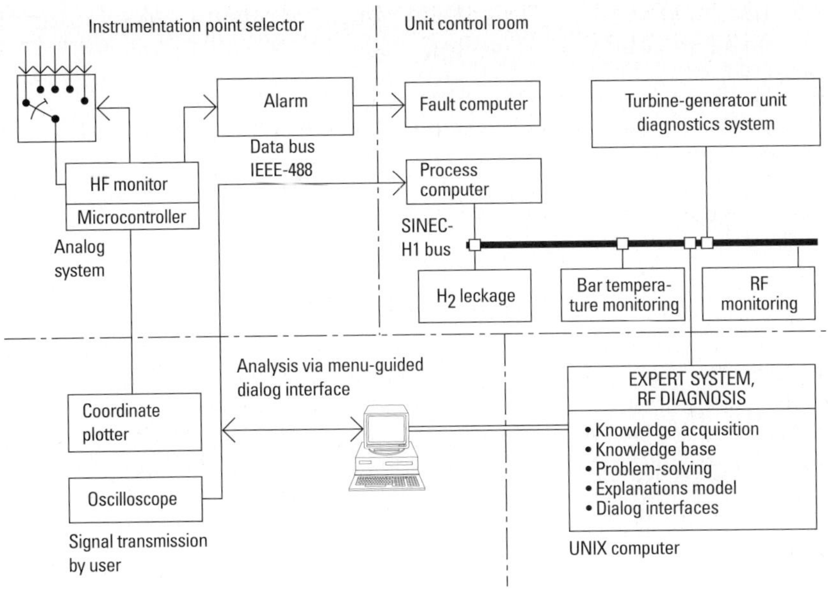

Figure 7.8.2.2 RF Diagnostics System, Extension Stage

verify a diagnosis and to run through alternatives. Furthermore, in Stage 2, he also has available numerous options for the visualization of measured data and characteristic variables in the form of graphics, diagrams or trend plots. This offers a qualitatively even better, reconstructible possibility for analyzing the results. The form in which the result of the diagnosis is presented is the same for both ways. A prerequisite for the analysis of the measured RF signals in the on-line mode with subsequent fully automatic diagnosis of the condition of the winding and insulation, is digital signal processing. Its technical implementation is achieved by way of three-stage computer-assisted processing:

- ▷ Signal preprocessing with analytical signal processing algorithms in a process microcomputer system,
- ▷ Calculation and analysis of RF signal characteristics,
- ▷ Transmission of the characteristic variables to the RF diagnostics.

The software includes the application of programs for measured data preprocessing and for the main process. For stand-alone application the software of the processing computer can run on all computers (independent of manufacturer) operating system.

7.9 Diagnostics Module for Detecting Hydrogen Leaks on Generators (H_2-Sealing)

In gas-cooled generators, "H_2-Safety" is of paramount importance. The "H_2-Sealing module" analyzes hydrogen gas consumption, distinguishing normal "breathing" from leaks. Data from the auxiliary systems make it possible to draw conclusions for pinpointing leaks. The module enhances safety against undetected, uncontrolled losses and diagnoses premature wear to the sealing elements [9,10].

Task of the Module

The hydrogen-sealing module has the task of monitoring the gas consumption and detecting slowly developing generator H_2 leaks and any malfunctioning of the generator shaft seals as well as to give an approximate location of the leaks. The continuous monitoring of about fifty significant measured variables and gradients is mandatory for the diagnostics function. The module configuration and function explained in the following description on the generator THFF type can also be used for different types of hydrogen-cooled generators.

7.9.1 Structure of the System

As an example, in generators with large ratings THFF type, water (H_2O) and hydrogen (H_2) are the media used to remove the heat produced. The cooling water (primary water) flows through the copper stator winding, rotor winding and circuit ring. Hydrogen at positive pressure is enclosed within the frame of the generator. The heat generated by the friction of the gas on the moving rotor, by hysteresis, and by eddy currents is absorbed by the hydrogen, which circulates around the stator frame at a pressure of, for example, 4 bar and transfers the heat to the secondary cooling water loop via coolers within the stator frame. The only components of the H_2-cooling system which are located outside the stator frame are the supply, drying and monitoring facilities and the equipment required to maintain the pressure of the gas in the primary water storage tank and of the purging gas for the water supply box. The generator frame is sealed against the atmosphere at the rotating shaft by an oil-lubricated oil shaft seal. The gas supply meets the safety regulations of VDE 0350/Part 1 and the compressed gas ordinance and takes special account of the properties of hydrogen. Figure 7.9.1.0 (derived from a hardcopy) gives an overview of the hydrogen-sealing system.

The gas supply system permits filling, makeup and drainage of the generator frame. To meet the pertinent regulations concerning the handling of hydrogen, unchecked losses must be kept to a minimum without exceeding a narrowly defined framework. Hydrogen is consumed for normal fluctuation, hydrogen must be used for the various purging tasks and pressure maintenance in the primary water system to permit catalytic oxygen

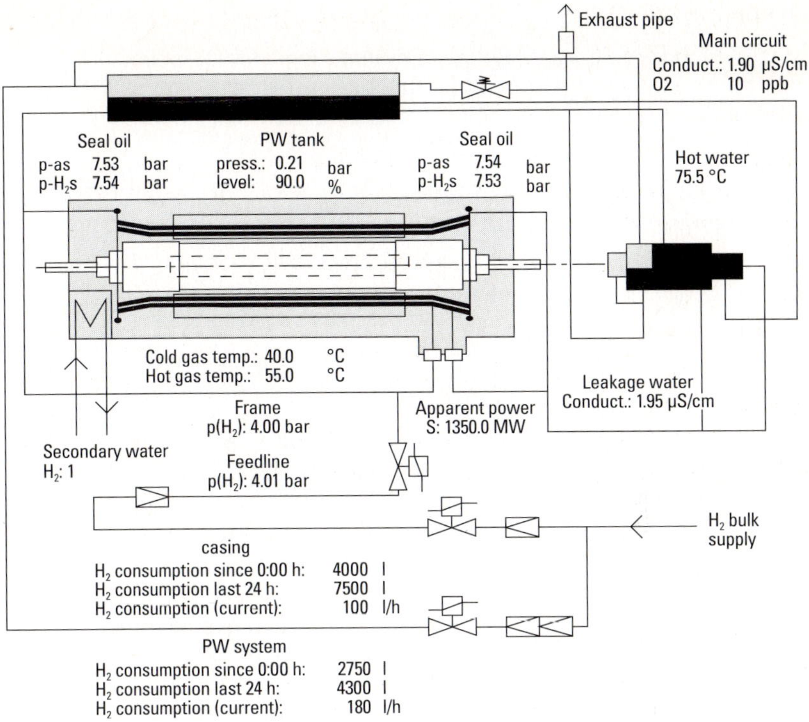

Figure 7.9.1.0
Display of Measured Data Relevant to Evaluation of the Hydrogen-Sealing System

reduction. The purging gas is fed into a chamber of the connection head that is subject to slight underpressure.

Acquisition and Preprocessing of Data

A SIMATIC S5 system is used for data acquisition and preprocessing. All the variables and status signals relevant to monitoring are recorded cyclically every second and converted into digital signals. The data acquisition computer also subjects the measured data to a plausibility check:

- ▷ Measured data not within the plausible measuring range,
- ▷ Non-coincidence of measured data,
- ▷ Gradient monitoring,
- ▷ Technological plausibility.

Only data which have been confirmed, are passed on to the diagnostics computer.

Monitoring

Using the monitoring and diagnostics system, H_2-sealing, incipient H_2 leaks from the generator and faults in the function of the generator shaft seal are analyzed and detected in good time. The system performs the following tasks:

▷ Monitoring and analysis of gas consumption,

▷ Early detection of hydrogen leaks,

▷ Approximate location of leaks.

Monitoring of the H_2 cooling system is used to detect possible leaks through which hydrogen could escape into the:

▷ Atmosphere,

▷ Primary water-cooling circuit,

▷ Secondary water-cooling circuit and

▷ Bearing compartment, if there is a fault in the shaft seal.

Any loss of hydrogen (unchecked and/or checked) causes changes in pressure and hydrogen make-up. A possible leak can therefore be detected by a change in pressure or hydrogen make-up. However, the monitoring function must take into account the fact that the H_2 pressure can change for other reasons. Further measures and symptoms must therefore be included in the diagnostics to detect leaks and the changes they cause reliably and in good time. The data acquisition system SIMATIC S5 and the monitoring of the hydrogen-cooling system are interconnected via a powerful bus. The division into two systems permits simultaneous execution of different tasks. This makes optimum monitoring for H_2 leaks possible.

If a fault is detected, a signal is outputted immediately. The status of a defective value is set to “BAD-VALUE” until the fault is remedied. This means that until the fault is remedied monitoring for H_2 leaks is only possible to a restricted degree or not at all. The measured data detected as being good are constantly stored in a short-term memory and their behavior is checked. For example: “normal or constant”, “increasing high”, “decreasing low”. The memory content is archived in a long-term memory in the down-circuit monitoring computer along with the signals about the behavior of the measured data and processed to form diagnostics on H_2 leaks.

7.9.2 Analysis and Diagnosis

In view of the need to build up a knowledge base step by step and modify and expand this knowledge base during operation, the decision to implement the diagnostics in the form of an expert system is logical. The requirements to be met by the system and the representation and processing of the diagnostics knowledge were defined by experts. This quotation contains the following two points of major interest: The diagnostics logic is represented in the form of rules; the variables required for diagnostics purposes are not processed directly in the rules, in the form of numerical values but are converted into natural-language ("linguistic") expressions via membership diagrams. An example for such a rule is as follows:

> *"If the H_2 consumption is high and the apparent power is not low and the hot-gas temperature is low or falling and the H_2 pressure in the generator frame is not low or falling and the cold gas temperature is falling or low, then the increase in H_2 consumption can be attributed to a decrease in cold-gas temperature".*

The logic gatings are shown in Figure 7.9.2.0. The approach taken toward the diagnostics is therefore in some ways already "fuzzy-like", for it permitted a problem-oriented formulation of the diagnostics report unencumbered by the concrete numerical values of the variables involved. The knowledge base is integrated into the general DIGEST operator interface.

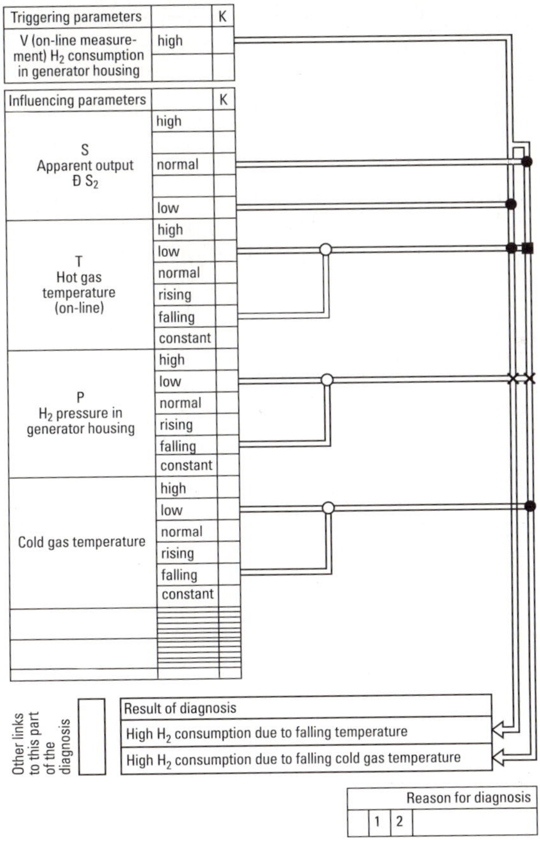

Figure 7.9.2.0 Hydrogen-Sealing, Diagnosis

How the System Works

The measured values are acquired cyclically and processed by the program in four steps. In the data preprocessing step, validity checks are made, deviations from the setpoint are established and gradients and averages are calculated. The data analysis step translates the numerically measured variables into linguistic statements (fuzzification). Overlapping of the membership functions means that one value may have more than one corresponding linguistic term. The generation of relevant intermediate analyses is implemented by mathematical links via defined operators. These represent specific internal states of the cooling system which cannot be observed directly from outside. Intermediate results can therefore be used as symptoms in other rules, i.e. the rules constitute a cycle-free network. The intermediate analyses are combined with other input data to determine the final result. The degree of fulfillment of the diagnosis is established in the process. Since the diagnoses are set statements and not linguistic variables, no defuzzification is necessary. The result is a list of diagnoses sorted according to their degree of fulfillment. How the process is performed is depicted in Figure 7.9.2.1.

Fuzzy Logic Elements

Representations consisting of two to five piecewise linear functions were chosen for the membership functions. It is generally sufficient to define three membership functions: As a rule, these are referred to as “low”, “normal” and “high” for normal values and “falling”, “constant” and “rising” for gradients. The standard function of minimum, maximum and complement were also used for processing the standard fuzzy operators. Because of their limited transparency and verifiability, compensatory operators were not considered. However, it soon became apparent that the standard operators were incapable of formulating the diagnostics knowledge adequately. Situations arose in which the existence of a symptom intensified and/or weakened the diagnosis, whereas its non-existence had no effect on the diagnosis. It was therefore decided to introduce two asymmetrical operators, an intensifying and a weakening operator. The mathematical conversion of the most important operators used in the fuzzy logic-expert systems is shown in the following:

and	(●)	min (a, b)
or	(o)	max (a, b)
not	(x)	$1 - a$
Intensifying	■	min $(1, a + (a \cdot b)/2)$
Weakening	□	max $(0, a - (a \cdot b)/2)$

On demand, further operators can be introduced and are already implemented in the mean time. As already mentioned, defuzzification is not necessary. The degree of fulfillment calculated by the fuzzy system is assigned to the diagnosis as confidence.

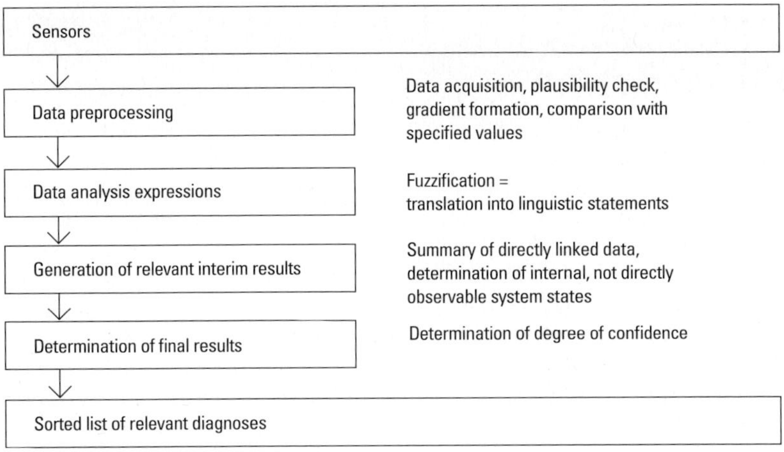

Figure 7.9.2.1 Function of the Fuzzy System

Why is Fuzzy Logic Used?

▷ The inputs are linguistic terms whose fuzzification has already been thought out by experts in the field,

▷ The logic links between the input data and the diagnoses cannot be represented solely by standard AND and OR gates,

▷ All data required are already available as sensor readings or outputs from previous processing stages, i.e. no query/reply cycles necessary,

▷ Graphic representation of related functions is the way experts think; this makes it more acceptable to expert users.

Knowledge Base

About 20 measured values are collected direct from the process in one-minute cycles; another 15 parameters are stored statically in the system. About 400 formulas for measured-value processing are used to generate about 60 input parameters for the fuzzy system. The monitoring logic, which decides whether or not to initiate diagnostics, has 14 trigger rules. The membership function yield about 130 symptoms, which are processed with about 110 rules and generate about 40 diagnoses. Although the hydrogen sealing module is doubtless one of the larger fuzzy expert system in existence, there are absolutely no performance problems when operating on a workstation. In the monitoring computer (workstation), irregularities in measured data are detected, classified and evaluated. If the irregularities are significant enough, a diagnosis is initiated. The result first appears in an abbreviated form on the screen with additional information about the probability of the information (degree of confidence). Windows with the following contents can be opened on the screen using operating elements:

- ▷ Detailed diagnosis text with suggestions for actions or additional tests,
- ▷ Reason for the diagnosis, explanation component with representation of the relevant rule tree,
- ▷ Diagrams and graphics of the necessary analyses that provide information about the behavior,
- ▷ Plant schematic with current measured data (see Figure 7.9.1.0)

The initiation of a diagnosis is event-oriented, where the definition of an initiating event may cover not only measured data which exceed the scatter band for the parameter in question, but also any other combination of factors, however complex. The signals issued after data preprocessing are archived in the long-term memory of the decentralized processing and monitoring computer, where they are checked for anomalies, classified and evaluated. If a sufficiently significant anomaly is established, such as: e.g. "changed high", "changed low", "rising", "falling". Each diagnosis consists of a written diagnosis, an assessment, reasons for the diagnosis and proposals for remedial measures or additional inspections.

On demand, a graphic display can be activated to illustrate how the diagnosis was derived from the data base of expert experience. For evaluation of the diagnosis, the user has at his disposal a variety of tools such as the representation of current measured data and the display of long-term trends (tables and diagrams).

The H_2-Sealing module can be used both individually (autonomous solution) and together with other DIGEST diagnostics modules. If the individual solution is used, the module has its own measured data preprocessing and monitoring computer. If a sum solution is selected, the module uses the measured-data preprocessing and the monitoring computer of the other diagnostics modules.

7.10 Diagnostics Module for Recognizing High Humidity-Conditions and Water Losses in Generators

Water entering the gas-filled generator raises humidity or may even wet components, reducing insulation capacity or triggering corrosion. Information link-ups make it possible to detect leaks via changes in the humidity of the gas and the duty cycle of the gas drier long before water loss from a cooling circuit is detected. Leakage rates are determined and changes recognized. In most cases, broad localization is possible, otherwise further action is recommended for more detailed investigation and to shorten the time taken to locate and repair the leak. Hazardous operating conditions are annunciated, depending on the location of the leak, the leak rate, and the duration of exposure to the water.

Task of the Module

This high humidity-conditions monitoring and diagnostics module monitors the gas humidity status in the generator. It recognizes abnormal ingress of moisture and performs diagnoses that make it possible to prevent hazardous operating conditions arising due to wetting of components. The module performs the following tasks:

- ▷ Monitoring and analysis of the gas humidity in the generator,
- ▷ Control of the gas driers (regeneration, operation),
- ▷ Measurement of gas drier duty cycles,
- ▷ Early recognition of abnormal water ingress,
- ▷ Diagnostics and trouble shouting.

7.10.1 Structure of the System (Example THFF Series)

The heat losses in high-output generators are removed via the cooling media water (H_2O) and hydrogen (H_2). Figure 7.10.1.0 shows the system diagram of a generator with water-cooled rotor windings. The water flows through the copper conductors of the stator winding, rotor winding and circuit ring, removing the heat losses. The hydrogen, which fills the entire generator housing under a considerable pressure, swirls around the turning rotor, the

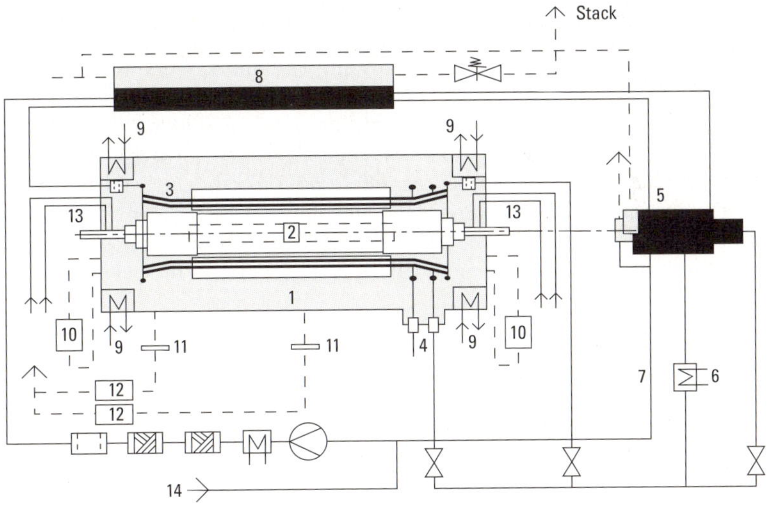

1 Hydrogen-cooled generator housing
2 Rotor winding, water-cooled
3 Stator winding, water-cooled
4 Circuit ring, water-cooled
5 Cooling water connection head
6 Primary water cooler
7 Leakage water circuit
8 Primary water tank
9 Hydrogen cooler
10 Gas drier
11 Humidity monitor
12 Purity monitor
13 Shaft seal
14 Demineralized water make-up

Figure 7.10.1.0 Cooling Circuit of the THFF Generator

stator core, and the support pieces. It removes the heat losses due to gas friction, magnetization and eddy currents. The hydrogen is recirculated within the generator housing and cooled via the heat exchangers located in the generator. The cooling circuit is a closed loop, enclosed within the generator housing.

The rest of the system consists of peripheral equipment for supplying, drying and monitoring the hydrogen. The generator housing is sealed to atmosphere by means of an oil-lubricated, oil-filled radial seal on the rotating shaft. The out-flowing fluid (water, oil) in case of malfunctions collected at the base of the casing, is monitored by liquid-level detectors. Evaporating water increases the "gas humidity" and can be detected via humidity sensors. Evaporated water is removed from the gas as it passes through gas driers filled with desiccant gel. The gel has a limited water-absorbing capacity. In this operating phase, the mean humidity in the gas drops sharply. After only a few hours, steam slippage becomes much higher until approximate equilibrium is reached between the moisture being entrained into the gas-filled space and the moisture being removed via the gas driers. In this operating phase, the mean humidity rises (more or less) slowly.

The period between the time when the regenerated drier is put on line and the time at which a certain humidity is attained, is known as the duty cycle of the drier. It is detected by the module's software and compared with previous duty cycles to recognize in good time any abnormal ingress of water into the gas-filled space. At the end of each duty cycle, the desiccant gel fill is exhausted and has to be regenerated, i.e. dried in hot air. The analysis identifies when changeover (duty-regeneration-duty) is necessary. The absorption capacity of absorbers is a function of the relative humidity, i.e. it depends on the steam temperature and pressure. The absorption capacity is greater at high relative humidities. The absorbers in the gas circuit are not only the desiccant gel fills but also the insulation. The insulating materials absorb moisture from the atmosphere when the generator is open during installation or servicing and release it again into the dry hydrogen fill, as a function of temperature, when the generator is in service.

For this reason, it is difficult to distinguish between moisture in-leakage and moisture due to evaporation from the insulation in the early phase of operation after initial installation or shut-downs for servicing. However, both can increase gas humidity to such an extent that the dewing condition is reached which may lead to corrosion sets in at the heat exchangers and (rarely) at components exposed to primary water flow. Loss of water from the primary water system can lead to the generator being shut-down due to inadequate cooling. The module monitors the humidity of the hydrogen fill, the duty cycle of the gas driers, and the water inventory in the primary water system. Comparison with relevant reference values from normal operation enables conclusions, to be drawn as to potentially fault-induced changes and indications, to be given in the form of diagnoses in time to avert hazardous operating conditions.

Data Acquisition and Conditioning

The SIMATIC S5-process monitors scan all analog and binary input parameters in a one-second cycle, convert them into digital measurement signals, check them for (physically) plausible values, and compress them to one-minute averages. These values are coded into telegrams for transmission to the monitoring computer. To prevent misinterpretation, a comprehensive plausibility check is provided for the humidity measurements. The measurement signals of significance to the humidity problem area are read in at short cyclic intervals and compared with their standard value and with past (historic) values. Measurement signals that are found to be faulted are reported, their statuses set to "BAD VALUE" pending elimination of the fault, and are ignored for the purposes of the analysis. The monitoring (and diagnostics) computer uses these measurement signals to derive symptoms which it stores together with the associated measurement signals in the short-term memory and monitors for their further development, e.g.:

- ▷ Normal and constant,
- ▷ High and changing,
- ▷ Low and changing.

The measurement signals and the results are archived in the long-term memory and are processed in the diagnoses performed by the module. Important general data that are not available from the system can be read in by hand, if the module is used as a stand-alone module.

7.10.2 Analyses and Diagnoses

The module analyzes the gas humidity and the duty and regeneration cycles of the gas driers and generates diagnoses and instructions to the control room personnel to take actions such as:

- ▷ Changeover of gas driers to regeneration mode or,
- ▷ Change back to duty mode.

Irregularities in the measurement signals are recognized, classified and recorded, and diagnoses are generated. The diagnoses are based on the expert knowledge stored in the knowledge base and give indications on actions to be taken to avert hazardous operating conditions, such as can occur if wetted components such as the stator core, the rotor body or, in the long term, even the rotor retaining rings corrode or if the dielectric strength of the insulating materials deteriorates. The knowledge base contains evaluation criteria for each symptom and rules in which the qualified symptoms are linked up. The output of the rules is one or more diagnoses, each with its associated confidence factor. The abbreviated text of the diagnosis is always printed out immediately via a data printer, and a parallel message can be sent to the control room. If the user interface is activated (i.e. the computer is on-line), the incoming diagnosis is first shown on screen together with its confidence factor. The user can then open up further windows with the following contents:

- ▷ Detailed diagnosis with proposals for remedial action or further investigation,
- ▷ Diagnosis trace in graphic form to show how and from which symptoms the diagnosis was derived,
- ▷ Freely programmable graphic displays showing measurements from the past twelve months.

The knowledge base implemented in the module is flexible, i.e. new findings can be incorporated at any time. In the stand-alone version, this diagnostics module has its own measurement signal preprocessing unit and its own monitoring computer. In combination with other modules, it shares the facilities with these.

7.11 Diagnostics Module for Recognizing Disturbed Cooling of Generators with Water-Cooled Rotor and Stator

The cooling medium in the cooling systems removes the heat losses from the generator to keep heat-up of the windings and their components within allowable limits. As a rule, a fault in the cooling system will cause overheating and destruction of the insulating materials, occasionally even melting of active metal parts. In addition to the generator condition monitor for stator bar temperature monitoring for water-cooled stator windings, this module has been developed for diagnosis of the water cooled rotor and stator windings and primary water system of the type THFF generator.

Task of the Module

The task of this module is to detect and pinpoint zones with disturbed rotor and stator cooling and disturbances in the primary cooling-water circuit (rotor and stator and current bushings and circuit rings). Action can be taken to prevent the damage spreading, and in most cases it will even be possible to continue plant operation at a suitably adapted output level. The monitored parameters, the analyses and diagnoses performed are:

- ▷ Monitoring and analysis of pressures, volumetric flow rates and temperature rises,
- ▷ Early indications of flow restrictions,
- ▷ Analysis of the cooling water chemistry,
- ▷ Diagnoses and support by fault analysis and countermeasures,
- ▷ Visualization of actual and measured data by trend plots and with the freely configurable graphics.

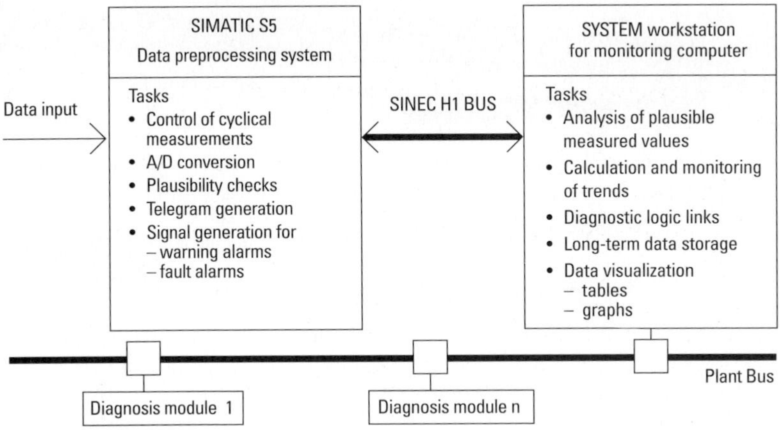

Figure 7.11.0 Data Processing and Objective of the Module

An overview of the data processing and objective of the system is given in Figure 7.11.0.

7.11.1 Structure of the System (THFF Series)

The primary water system is a closed loop in which the water is recirculated, cooled and conditioned to ensure a constant water quality. A storage tank (1) is located at the highest point in the system (see Figure 7.11.1.0), from which the water flows to the pump (2), which here forms part of the primary water supply head. This primary water supply head (box) encloses the end of the shaft on which the impeller wheel of the pump is mounted. From the pump, the water is passed through the heat exchanger assembly (3) to the filter (4). Downstream of the filter, the water flow splits and passes through the circuit ring loop (5), the stator loop (6) and the rotor loop (7). A fine-mesh filter is located upstream of the rotor. In the primary water supply box, the water to the rotor is pushed into a hole at the center of the end face of the shaft and then distributed to the copper conductors of the rotor winding, which have a axial hole to guide the water. The heated water leaving the rotor is collected in a chamber of the primary water supply box and returned to the storage tank.

The module monitors the subloop and recognizes obstructions of flow in the outer loop and in the turning rotor. It compares the instantaneous operating point (*delta p*, *v*) with the hydraulic characteristics of the undisturbed system and reports any changes early and in good time. All seals between the primary water supply box and the rotating shaft are non-contacting split-bushing-type labyrinth seals across which water is exchanged. Water flowing out to atmosphere is passed via the leakage water circuit (9), in which any oxygen, carbon dioxide and dissociated substances are removed, back to the main

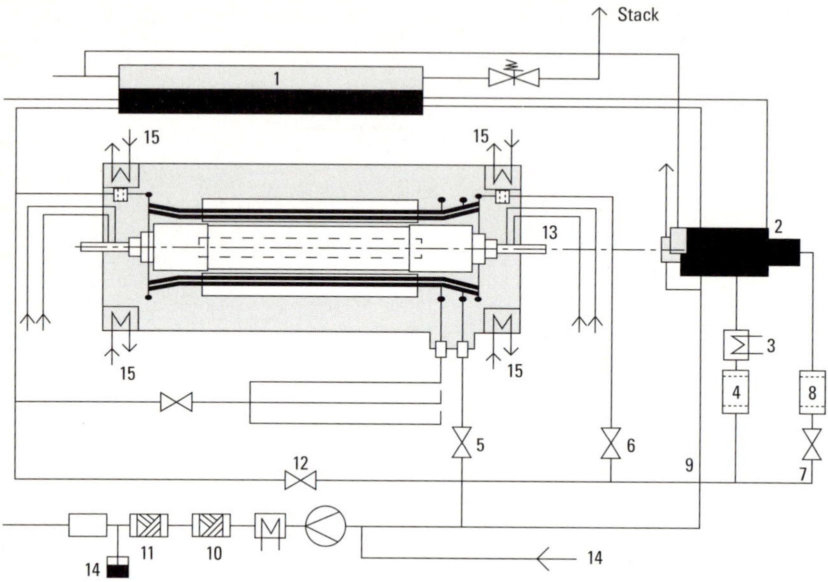

1 Primary water tank
2 Cooling water connection head
3 Primary water cooler
4 Primary water filter
5 Circuit loop ring
6 Stator loop
7 Rotor loop
8 Rotor loop filter
9 Leakage water circuit
10 Palladium catalytic reactor
11 Ion exchanger
12 Bypass
13 Demineralized water make-up
14 Alkalizer unit
15 Hydrogen cooler

Figure 7.11.1.0 Primary Water-Cooling Circuit of a THFF Generator

circuit. Damage to the split-bushing-type labyrinth seals, changes the amount of water crossing the seals and the pressures in the chambers of the primary water supply box.

The module recognizes these changes and generates diagnoses accordingly. In the stator loop, the water is passed through strainers, at the inlet to and outlet from the stator, which prevent course dirt from finding its way into the headers (during and in the period after installation). The stator conductor to be cooled is divided into strands. The strands discharge into water boxes that are connected to the headers via insulating flexible lines.

The generator condition monitoring module for the water cooled stator windings and this module complement each other in monitoring the stator loop. The first one recognizes obstructions in individual hollow conductor strands, while this module recognizes the shift in the hydraulic operating point of the winding as a whole. Monitoring the operating point makes it possible to draw conclusions and provides diagnoses as to whether a recognized change in the volumetric flow rate is due to a fault in the winding or in the outer circuit. In the circuit ring loop there are 2 current bushings with (each up to 4) parallel sections of circuit rings which are connected in hydraulic series. Each circuit ring section has symmetrical flow restrictors to ensure a uniform flow distribution. The three loops U-X, V-Y and W-Z are subject to the same pressure difference. A change in the volumetric flow rate in any one loop is indicative of internal faults, a change in all three loops indicates an external fault. A drop in the volumetric flow rate in combination with pressure surges is indicative of water boiling in the circuit. This module identifies these correlations and generates a diagnosis accordingly.

All water-cooled parts of the winding are made of copper, which corrodes and forms corrosion products in contact with water. These corrosion products form deposits in the filters and flow paths, and in hollow conductors, can create obstacles to through flow. The intensity of corrosion is minimized by appropriate water-chemistry management (10, 11, 13, 14). The waterchemistry parameters are likewise monitored. In the event of any changes in the water chemistry, the diagnostics module draws conclusions as to the possible causes and suggests further investigations as appropriate. Similar changes in the hydraulic conditions in all cooling loops have their origins in that part of the system from which all loops are fed, whereas changes in only one loop can only be due to causes within that loop. The module investigates these causes and is capable of localizing the region affected and possibly even of identifying the fault, e.g. operator error, a clogged filter, or a defect in the shaft-driven pump.

It will not always be possible to eliminate the diagnosed fault while the generator is in service; on the other hand, it will not always be necessary to eliminate the fault immediately. The important thing is that the fault has been recognized. Remedial action can then be taken at the next opportunity or during a scheduled inspection shut-down.

Data Acquisition and Conditioning

The SIMATIC S5-process monitors scan all analog and binary input parameters in a one-second or two-second cycle, convert them into digital measurement signals, and compress them to 1-minute averages. These values are checked for plausibility and coded into telegrams for transmission to the monitoring computer. Further plausibility checks are performed in the course of the analysis.

Measurement signals that are found to be faulted are reported, reset to "BAD VALUE" pending elimination of the fault, and are ignored for the purposes of the analysis. The monitoring (and diagnostics) computer uses these measurement signals to derive symptoms and stores these together with the associated measurement signals in the short-term memory and monitors for their further development, e.g.: "normal and constant, high and changing, low and changing".

The measurement signals and the results are archived in the long-term memory and are processed in diagnoses, performed by the module. If, in any plant, not all measurements are present in the form of analog signals, these can be input in the above cycle. For this reason, provision is made for values to be read off and input during a tour of inspection. These are coupled with the on-line data and evaluated, and make further diagnoses possible.

7.11.2 Analyses and Diagnoses

The module monitors the entire primary water system, it recognizes and evaluates irregularities in measurement signals and generates diagnoses as soon as the irregularities go beyond the specified tolerance range and the derived symptoms indicate a fault with adequate certainty. The diagnoses are based on the expert knowledge stored in the knowledge base. This contains evaluation curves for each symptom and rules in which the qualified symptoms are linked up. The output of the rules is one or more diagnoses, each with its associated confidence factor. The abbreviated text of the diagnosis is always printed out immediately via a data printer, and a parallel message can be sent to the control room. The module reads in all significant measurement signals from the cooling system at short, cyclic intervals and evaluates them by comparison with the normal reference value and previous values. If the user interface of the module is activated (i.e. the user is logged in and the module has been booted), the incoming diagnosis is first shown together with its confidence factor on screen. The user can then open up further windows with the following contents:

- ▷ Detailed diagnosis with proposals for remedial action or further investigation,
- ▷ Diagnosis trace in graphic form to show how and from which symptoms the diagnosis was derived,
- ▷ Freely programmable graphic displays showing measurements from the past twelve months.

The module can be used separately (stand-alone version) or together with other diagnostics modules from the DIGEST family. In the stand-alone version, this diagnostics module has its own data acquisition unit and its own monitoring computer. In combination with other modules, it shares the facilities with these modules.

7.12 Gas Turbine Thermodynamics and Fault Analysis Module

The gas turbine module supplements existing operational instrumentation and control, but functions largely independently so that even time-consuming tasks such as complex thermodynamic calculations can be performed. In addition, analyses that have no direct impact on plant operation, but are of long-term interest, can also be performed. It analyzes deviations from specified values occurring during operation of the gas turbine. Analysis of the important operating data at various load conditions makes it possible to recognize discrepancies in power output, efficiency and incipient mechanical changes in running performance and to pre-empt incipient faults in the gas turbine [27, 28].

Task of the Module

The task of the gas turbine module is to perform comprehensive thermodynamic calculations. Actual values converted to "standard ambient conditions" provide detailed information for ascertaining the gas turbines condition. This process enables measurements taken under the most diverse conditions to be compared and differences between the current turbine status and the reference status to be shown. Transient conditions are evaluated, the respective data stored for the important operating processes and allow an evaluation of the turbine condition. Figure 7.12.0 shows an overall picture of the module. The gas turbine analysis module consists of various task-specific submodules that are combined to meet the individual need of the specific plant:

- ▷ Thermodynamics,
- ▷ Analysis for early detection and recognition of faults.

The selected modules have a common data acquisition section and user interface.

7.12.1 Analyses and Thermodynamic Calculations

Analysis of Steady-State Conditions

When analyzing a gas turbine operating under steady-state conditions, certain measured variables are used in an evaluation model to determine operating characteristics, which are then compared with desired values. The accuracy of this type of analysis is decisively dependent on the parameters selected. The objective is to recognize changes as early as possible for alarm annunciation, thus enabling the operator to react quickly. For example, the temperature distribution of the exhaust gases immediately downstream of the turbine is analyzed. Output, fuel consumption, and thermal equilibrium have

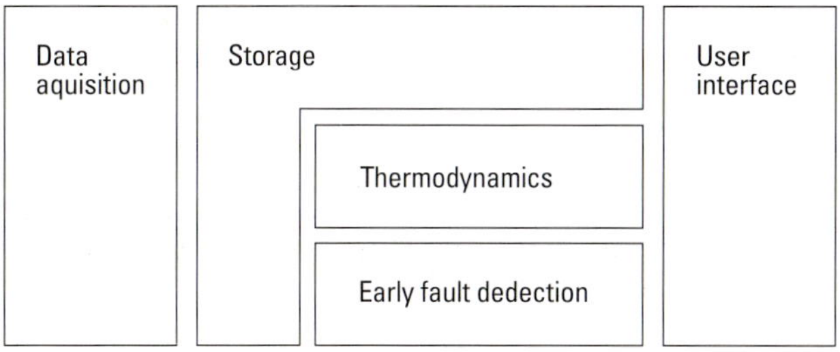

Figure 7.12.0 Structure of the Gas Turbine Module

proven to be decisive parameters. Changes in the temperature field can be determined while each individual value is still within permissible limits. Irregularities in the combustion system, such as leaks, can also be detected.

Analysis of Transient Conditions

In transient condition analysis, the gas turbine's response to a specific control signal is evaluated. In place of the classical step-function or ramp signals known from control engineering, gas turbines can be operated with combinations of input signals or defined, nonlinear parameters. For example, turbine speed is monitored during start-up or coast-down, triggered by a combination of different signals. Changes in coast-down behavior, such as larger speed gradients, are indicative of malfunctions. For comparison purposes, data stored for all operating processes (e.g. during commissioning) can be displayed or superimposed for: Start-up, loading, unloading, coast-down curve, load rejection, turbine trip. In the event of a sudden turbine trip, current data is stored without any data reduction.

Identification of Model Parameters

The evaluation models used to determine characteristics utilize classifications as well as equations. The gas turbine module determines the necessary parameters during a learning phase, thus ensuring that turbine-specific factors are taken into consideration. Classifications are used in analyses in which the relationships cannot be adequately formulated in mathematical terms, such as the temperature distribution at the turbine exhaust. The model requires a large number of reference values and is only applicable to a narrow range in each case. Processes that can be described mathematically can be analyzed over a broad range with sufficient accuracy using only a few reference values. One example is the analysis of the coast-down behavior of the gas turbine.

Thermodynamic Calculations

If the gas turbine is running in a steady-state condition, comprehensive thermodynamic calculations are performed in cycles using the measured data. Actual values and reference values corrected for standard ambient conditions are determined. These data provide detailed information for ascertaining the gas turbine condition. With the thermodynamic module, important gas turbine process data can be monitored on-line, and inferences on the state of the gas turbine can be drawn from the comparison of various measurements. The monitored significant parameters of the thermodynamics module are as follows:

▷ Overall Plant
- Ambient conditions (air temperature, pressure, relative humidity),
- Output at terminals, auxiliary power load,
- Terminal efficiency, specific heat consumption,
- Speed.

▷ Compressor
  - Mass flow rate at intake and pressure ratio,
  - Efficiency (isentropic and polytropic),
  - Outlet temperature.

▷ Combustion System and Turbine
  - Inlet temperature (to ISO 2314), fuel flow rate, air-to-fuel ratio, mass flow rate and exhaust-gas temperature,
  - Pressure drop combustion system, pressure ratio turbine,
  - Efficiency turbine (isentropic and polytropic).

▷ Generator and Gear
  - Generator and gearing efficiency.

The central component of the thermodynamics module is a computer program which evaluates measurements and converts measurement results to conform the predefined boundary conditions. The thermodynamics module is capable of converting significant measured variables (shown in the top field of Table 7.12.1.0) into selectable reference values. Additional parameters can also be taken into consideration in two further steps (center and bottom fields). Alone, this program is also used for data evaluation. In this way, all data changes which affect the evaluation program, are automatically incorporated into the analysis version.

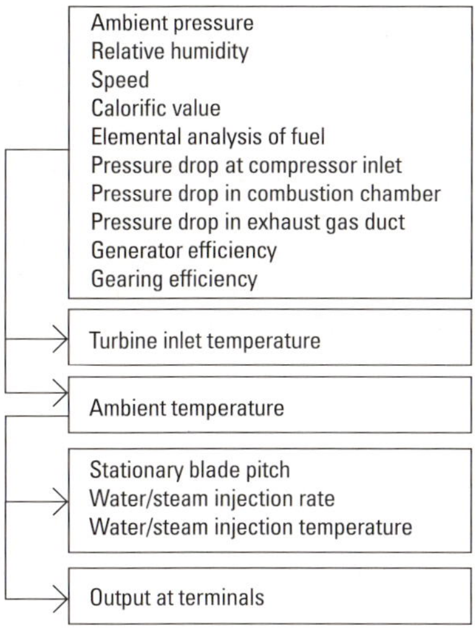

Figure 7.12.1.0
Thermodynamic Converting Characteristics for State Evaluation

When appropriately calibrated measuring equipment is utilized in the gas turbine, the module is capable of performing thermodynamic warranty acceptance measurements with no additional hardware or software. Comparison has shown that the precision of these measurements meets all requirements of warranty acceptance measurement.

Analysis of Measurements

The families of characteristics describing the operating behavior of the gas turbine under different ambient and load conditions are used to convert the current measuring point to user-specific reference conditions. This process enables measurements taken under the most diverse conditions to be compared, and differences between the current turbine status and that of a new, clean turbine to be determined, for example. This analysis is supported by the module's built-in capability to generate a time-dependent plot of the conversion results.

This process enables the operator to assess the effects of compressor and turbine aging and blade fouling, and select the most economically favorable time to clean the compressor blading or perform a major inspection and overhaul. The efficiency of new gas turbine decreases with time. A typical characteristic curve is shown in Figure 7.12.1.1, curve 1: Increasing blade surface roughness reduces gas turbine output to a certain extend. In addition, fouling of the compressor blades can considerably reduce efficiency. The output can be increased again by cleaning the compressor during power operation (curve 2) or at turning speed (curve 3). A certain amount of gas turbine efficiency is recoverable by a major inspection with blading overhaul.

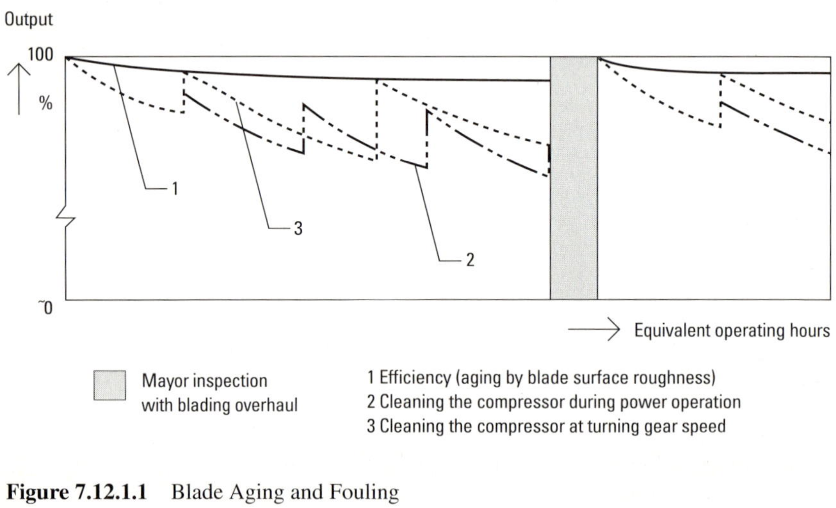

Figure 7.12.1.1 Blade Aging and Fouling

7.12.2 Data Processing and Hardware

Data Acquisition

Signals are input and converted, or adapted in accordance with set characteristics (e.g. from calibration) continuously. Acquisition can also be achieved using the HP 3852 S programmable data acquisition unit (Hewlett Packard) or through a SINEC H1 bus interface connection.

Measured Variables

The installed sensors and the data logger provide for example precise data on the following: turbine speed, electrical output, blade pitch adjustment, compressor inlet temperatures, pressure at compressor inlet, fuel mass flow as listed up before. Bearing temperatures, vibration of the bearing pedestals and shaft vibration can also be included as an additional option.

Recording the Equivalent Operating Hours (Incorporated in the I&C Equipment)

In Siemens/KWU gas turbines, the time intervals between major inspections and overhauls are determined based on equivalent operating hours. Equivalent operating hours reflect component loading conditions caused by the different modes of turbine operation (start-up, rapid load changes, base-load operation). The equivalent operating hours can therefore indicate the need for an inspection and overhaul before component material fatigue has reached a critical point. A service life expenditure model based on the actual strains sustained by the blading is currently under development.

Vibration

In addition to the gas turbine module, a shaft vibration analysis module is also available basically of the same structure as used for the steam turbine generator set.

For gas turbines that do not use a shaft vibration module, a "Balancing Submodule" for data transmission of turbine vibration data to Siemens/KWU is available. These data will be used for the calculation of balancing-planes setting to correct shaft imbalances. The calculation performed at Siemens/ KWU and the resulting balancing weights given to the customer, enable the operators to set the balancing weights.

Recorded Process Data

The system computer, which receives measured data and binary signals from the I&C equipment, continuously monitors the current operating condition. Depending on the operating condition, these data are conditioned, stored, calculations of key data performed and made available in graphic form: Temperatures, mass flow, compressor pressure ratio, output, efficiency, heat consumption, operation time.

Computer Program and Hardware

The computer program itself makes use of the generic physical principles of gas turbine operation. It accesses turbine-specific data records which are contained in a type file. In this file, the individual characteristic data and families of characteristics of the individual components such as the compressor, combustion system, turbine, and generator are stored numerically. The measured data are read by a PC continuously. The PC used in the system, is a high-performance computer.

7.13 Outlook of Anticipated Modules

The further development of diagnostics modules and the development of new modules will continue to call for intensive efforts to be made in the core fields of technical diagnostics in the future, too. Particular attention is currently being paid to on-line diagnostics which makes it possible to make verbal statements as to incipient disturbances and to back up these verbal statements with eye-catching graphic displays. Modules to be developed are briefly described, some of them are under development and/or approaching completion.

7.13.1 Bearings

Extensive instrumentation is required for detailed representation of the momentary status of the bearings oil supply shaft lift function group. Even slight changes in certain parameters trigger comparison with documented standard statuses, an analysis of the peripheral conditions, and as precise as possible identification of the background causing the changes. A rise in bearing temperatures, for example, may be the result of misalignment, inadequate oil supply, overload on the bearings, or other influences. Early detection of disturbances, timely recommendation of suitable counter measures, and the avoidance of hazardous conditions are some of the objectives fulfilled by diagnostics in the interest of plant availability. Comprehensive diagnostics will require use of the complete module group. This module is backed up by analysis of shifts at the foundation slab of the turbine generator set (turbine plate deformation).

Measurements can be performed (e.g. with a mobile measurement device) only when it is considered necessary. The change in the deflection curve is calculated from the results of the turbine plate deformation measurements and is used to determine the resulting loads on the mating shaft ends.

7.13.2 Turbine Casings

Further diagnostics tasks are the monitoring and assessment of casing distortion, including cold steam or water back-flow phenomena, and the control of condensate drain function as well as the proper function of the gland steam sealing. The casing distortion can be calculated on-line with the finite element method using temperature measurements at the casing. The casing distortion shows the influence of the radial clearance between stationary and rotating parts and provides messages to avoid rubbing.

7.13.3 Valves

In steam turbine installation diagnostics, further modules are being developed to analyze the steam turbine valves. The turbine complex consisting of open-loop and closed-loop control, valve actuation and operation is dealt within a dedicated module, with importance being attached to test sequences. Incorporated are the lift monitor for main steam, reheat steam and bypass valves. The module includes long-term memory for the test data and parameters.

7.13.4 Excitation

New, non-contact scanning and data transfer systems nowadays make it possible to access the on-line parameters of the rotating parts of the excitation system, significantly extending the scope of diagnostics in the excitation area. For example, response of any fusing element in the rotating rectifier is immediately detected and correlated with the exciter status, and the quality of trip-out is assessed. If several fusing elements have tripped out, the remaining current-carrying capacity of the exciter is calculated and shut-down or power reduction is recommended.

7.13.5 Feed Water Chemistry

Safety against corrosion and preservation of material integrity make stringent demands of the water-chemistry conditions in the water/steam cycle of power plants. Conventional monitoring of the water/steam cycle on the basis of the readings provided by the instrumentation systems and analysis of laboratory samples is not sufficient if the aim is to be able to recognize incipient disturbances in good time. The objective of this module is to generate a diagnosis via changes and perturbations in the water-chemistry conditions in the water/steam cycle. The system detects and analyzes the relevant operating parameters and uses them to derive reference values against which any changes can be compared and interpreted. The diagnoses are based on the expert knowledge stored in the knowledge base of the fuzzy logic expert system DIGEST and contains evaluation for each symptom.

The subjects of monitoring are the power plant components exposed to the influence of the water-chemistry conditions in the plant. For nuclear power plants, the module incorporates the specific knowledge to provide comprehensive analyses and diagnoses in the various fields of water chemistry. Provision of the expert knowledge and the upkeep of the system are the responsibility of the water chemistry department of Siemens/KWU. In the meantime, the development of this module has been completed.

8 Benefit of DIGEST

Proper power plant functioning, optimum operation in terms of thermodynamics and optimized maintenance are essential to cost effective power generation. With a view to achieving these aims, a high level of automation and a comprehensive instrumentation and control equipment complement are state-of-the-art in modern plants. Diagnostics modules for assessing the condition of turbine generator sets have been developed within the scope of the DIGEST system. These modules are already proving to be helpful tools for further improving the efficiency and economy of power plants.

Modern diagnostics systems establish the link between instrumentation technology and data processing. New measurement techniques and modern sensors make it possible to keep track of process and component parameters that just a few years ago would have been unimaginable as reference variables [29]. For instance, it is now possible to measure the vibration of LP rotor blades on-line. Many more such examples could be given. Improvements in data processing technology in recent years, and in particular falling prices for increasingly powerful hardware, have set the stage for this development. This is what created the basis for the employment of diagnostics systems.

It became possible to develop programs which are capable of representing trends, storing and evaluating voluminous data records, and correlating various sequences. Once the allowable deviation of certain variables from their specified values has been quantified and certain constellations of measurement signals associated with known disturbances have been defined in the data processing part, constant comparison of the measured values of process variables with the stored data makes it possible to recognize incipient changes and disturbances and to adopt countermeasures in good time. Developments in the field of mechanical and electrical engineering have also made significant contributions to the ability to assess the condition of machines on-line and in shut-down conditions without having to open them for inspection.

There is thus no doubt whatever that modern diagnostics systems yield a broader overview of plant conditions than the monitoring systems used in the past. The question therefore arises as to in which fields the primary use of such diagnostics systems lies, the objectives of such applications and, above all, what benefits are to be gained through the use of diagnostics systems.

Primary fields of application and objectives imposed on diagnostic systems focus on the following:

Thermodynamically optimum mode of operation	⇒	Operation of plant with greatest possible efficiency
Detection of incipient faults and failures	⇒	Avoidance of sudden failures and secondary damage and production bottlenecks
Monitoring of service life expenditure	⇒	Flexible operation of the plant for maximum service life
Status-oriented maintenance strategy	⇒	Avoidance of scheduled outages (non-availabilities) for overhaul of the turbine generator set

Further advantages of the use of diagnostics systems ensure from the following effects, which can be assessed on a plant-by-plant basis only:

- ▷ Savings in personnel thanks to continuous optimization of process monitoring,
- ▷ Systematic collection of knowledge for better recognition of disturbances,
- ▷ Reduction in insurance premiums by avoiding secondary damage,
- ▷ Lower emissions to the environment due to thermodynamically optimum operation.

Table 8.0 Calculation of Cost Savings for a 400 MW Fossil-Fired Power Plant

Potential savings	Calculation method
Cost savings for a major inspection	By experience 2 - 3 Mio DM
Cost of repairing major damage	By experience 0.5 - 5 Mio DM
Power station outage of 4 weeks	$\frac{\text{Power} \times \text{specific costs} \times \text{annuity} \times \text{erection time factor}}{\text{12 months}}$ (400.000 kW × 2000 DM/kW × 0.1/year × 1.2/ 12 mont. ~ 8.0 Mio DM)
Purchase of electricy from external source of 4 weeks	Power × time × Δcosts × utilization factor (400.000 kW × 30 days × 24 h/day × 0.05 DM/kWh × 0.7 ~ 10 Mio DM)
Change of efficiency by 0.25 percentage points	$\frac{\text{Power} \times \text{service hours} \times \text{fuel costs}}{\text{Annuity}} \times \Delta\text{heat rate}$ (400.000 kW × 6100 h/year × 7/10^6 DM/GJ × 45 kJ/kWh/0.1/year ~ 7.6 Mio DM)
Change of availability by 0.5 percent	Power × specific costs × erection time factor × availability change (400.000 kW × 2000 DM/kW × 1.2 × $(1 - \frac{1}{1.005})$% ~ 4.8 Mio DM)

Quantification of the benefit which can be achieved through the use of diagnostics systems is only possible if the events of plant operation are foreseeable. Diagnostics systems are based on preventive operation. Their cost-effectiveness can therefore only be verified in the light of events that actually occur. It is a well-known and undisputed fact that prompt recognition of faults through the use of diagnostics systems can bring about substantial savings by prevention of consequential damage, hence the implementation of diagnostics systems is more than justified from an economic standpoint.

Experience has shown that decisions by plant operators on whether or not to implement diagnostics systems are often based on overly optimistic estimates of anticipated availability of plants and not on achievable savings in case of malfunctions and optimization by application of diagnostics. The actual achievable benefit of plants, of course, also depends on operator-specific factors. The task of demonstrating the savings which are possible through implementation of diagnostics systems must thus be considered in its range of the afore-mentioned application, substantiated by relevant experience gathered in applications.

Table 8.0 presents the calculation fundamentals for estimating the costs incurred by individual plant operating events. The magnitude of the resultant outage costs can be roughly estimated using these calculation parameters. Other overheads (e.g. personnel, maintenance and insurance costs) must also be allowed for independent operating events.

Table 8.1.0 Power Deficits to Design Parameters for Different Types of Power Plants

Construction parameter	Parameter change	Type of steam turbine 400 MW reheat Deviation in output MW	1300 MW saturated Deviation in output MW
Main steam temperatures	- 10 K	- 1.2	-
Main steam flow	- 0.5 %	- 2.0	- 6.0
Moisture down stream MS	+ 0.5 %	-	- 2
Hot reheat temperature	- 10 K	- 0.9	-
ReheaterTTD	+ 5 K	-	- 1.5
HP/IP efficiency	- 1 %	- 1.0 / - 1.0	- 5.0 / - 5.0
Averaged feedwater heaterTTD	+ 1 K	- 0.4	- 1.5
One feedwater heater bypassed	-	- 2.4	- 7.0
CondenserTTD	+ 1 K	- 0.6	- 2.0
Circulating water flow	- 10 %	- 0.6	- 2.0

8.1 Thermodynamically Optimum Mode of Operation

The high energy efficiency of today's power plants are accomplished by the use of complex cycles. Optimum thermodynamic operation is therefore an imperative to ensure economical operation of the unit. By constant comparison of actual plant conditions with design values, the thermo-dynamics module ensures optimum operation of the plant. Efficiency variations are noted and information on component performance is given.

Deviations of the operating parameters from their design values affect the heat rate: Table 8.1.0 shows such key parameters for a fossil-fired power plant and for a nuclear power plant and Table 8.1.1 for a combined-cycle power plant. Though it is theoretically possible for thermodynamic engineers to calculate the effects of such parameter discrepancies, continuous on-line performance of these calculations would require a major manpower effort. Continuous on-line diagnostics by the thermodynamics module, however, identifies the heat rate deficits even of parameters which require large-scale calculation to determine their effects on heat rate.

The evaluation of output deficits can be performed on a "cash-value" basis to show for typical power plant types how continuous analyses of the plant operating status by means of the thermodynamics module helps the plant operator to make savings in heat rate and to avoid output shortfalls. The savings can be calculated by the formula given in Table 8.0. The calculation by this method does not include a further, not inconsiderable loss due to reduced output that applies when the plants are operated at full load.

Table 8.1.1 Power Deficits to Design Parameters for a Combined-Cycle Power Plant

Design parameter	Parameter change	450 MW CCPP Unit Tripple-pressure, reheat Deviation in output MW
Gas turbine efficiency	1 % down	- 3.1
Compressor mass flow	1 % down	- 4.5
Compressor efficiency	1 % down	- 4.4
Main steam mass flow (at constant main steam temperature)	1 % down	- 1.5
Main steam temperature	10 K down	- 0.3
Hot reheat temperature	10 K down	- 0.3
HP/IP turbine efficiency	1 % down	- 0.3 / - 0.5
LP turbine efficiency	1 % down	- 0.6
Circulating water mass flow	10 % down	- 0.5
Terminal temperature difference condenser	1 K up	- 0.5

Analysis of operational deviations from the design parameters reveals that considerable savings can be made by adapting plant operation to process conditions even in the case of relatively minor deviations such as can be recognized by on-line diagnosis. An efficiency increase of the power plant results hence in relatively high savings. These cost savings finance a comprehensively equipped diagnostics system for a steam turbine unit or may even pay for one in full.

8.2 State-of-the-Art Methods for Fault Detection

The goal of obtaining detailed diagnoses for individual components requires the utilization of state-of-the-art methods and hardware. Table 8.2.0. shows a compilation of the methods and tools that are used today for on-line and

Table 8.2.0
Methods and Tools for the Status Monitoring of Steam Turbines and Generators for Off- and On-Line Diagnosis

Method	Diagnostic task (Module)	Frequency range accuracy
Magnetic tests	Generator module electromagnetic comp	Off-line, 50/60 Hz
Frequency analysis	Vibration module, torsion module, blade vibration	On-line, $10 - 10^3$ Hz $10^2 - 10^4$ Hz
Laser beam	Gap measuring module, blade vibration, foundation shifting	On-line, up to 10^4 Hz
Acoustic inspection	Leak monitoring (microphone)	Off-line, $10^3 - 10^4$ Hz
Eddy current inspection	Integrity of heat exchanger	Off-line, $10^3 - 10^5$ Hz
Radio frequency analysis	RF monitoring module for high voltage components	On-line, $10^5 - 10^7$ Hz
Electrical tests	Generator insulation status	Off-line, up to 10^7 Hz
Ultrasonics solid-borne noise examinations	Integrity of components	Off-line, $10^6 - 10^7$ Hz On-line, $10^5 - 10^8$ Hz
Infra-red photography	Changes in heat-up	On-line, $10^{12} - 10^{17}$ Hz
X-ray inspection	Integrity of	Off-line, $10^{17} - 10^{19}$ Hz
Borescopy video, CAT	Visual inspection, computer-assisted defect recognition of components	Off-line, image resolution down to 50 mm
Chemical analysis (ion select. method)	Oxygen, sodium	On-line, 0.2 to 1 ppb
Helium test with mass spectrometer	Leakages of heat exchangers and tubing	On and off-line, leaks > 0.03 mm diameter
Particle and operating media analysis	Lubricant	On and off-line

off-line diagnostics on steam turbine units. These methods and tools utilize the available technical possibilities and they cover the entire electro-magnetic spectrum, up to X-rays. Of course, there are a lot of other routine checks performed in off-line condition which are not listed here. Just to mention some sophisticated measurement techniques: Even the lowest concentration of the constituents of a medium can be determined using trace analyzers. Helium tests permit detection of the most minute leaks. Further to these measurements and analyses, a number of material tests and inspections, leak tightness tests, clearance checks, and deformation measurements are routinely performed.

Some of the measurements required for the generation of diagnoses cannot be performed while the turbine generator set is in operation. In this case, the measurements are performed when the machine is shut-down (off-line), but without having to open the casings. The results of off-line diagnoses in most cases make it possible to draw conclusion as to the condition of the component in question but can also provide other necessary information, e.g. as a contribution to the analysis of the disturbance or to the diagnosis. The DIGEST system provides for an "interactive dialog" with modules that produce a diagnosis, and the results gained from off-line investigation can be incorporated into the system to add to the symptoms that trigger the diagnostics runs.

The off-line diagnosis will be supported by inspections using modern testing techniques. Even today, turbine generator sets are already being designed and equipped to permit the performance of large-scale measurements on the

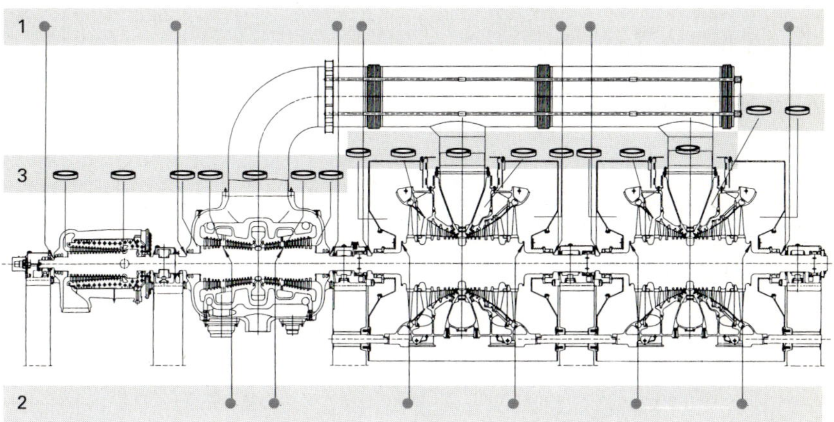

1. Location for balancing accessible without any dismantling
2. Location for balancing only accessible after dismantling
3. Borescope ports

Figure 8.2.1 Steam Turbine (HMN Series), Locations for Shaft Balancing

unopened machine or to obtain information on the internal condition of the machine simply by opening readily accessible, relatively small covers or manholes. Great importance is given to accessibility of the main components (valves, casings, bearings, etc.) and rapid performance of measurements features to permit these are being incorporated. One example shows how readily accessible openings on the turbine generator set can be used for inspecting the internal condition of the machine by performing borescopic examinations during shut-down (Figure 8.2.1). Access to the low-pressure turbine, the condensers and the generator is also possible via manholes, while facilities for re-balancing the turbine are provided at readily accessible external locations, via removable covers and, of course, by the possibility of installing further balancing weights in the turbine interior. Special-purpose measurements can be performed by our laboratory personnel using the above-mentioned modern measurement techniques.

An example of a modern inspection method is shown in Figure 8.2.2, which depicts how remote visual inspections can be performed on turbine blading. Lighting the inner side of the casing enables certain blades to be checked

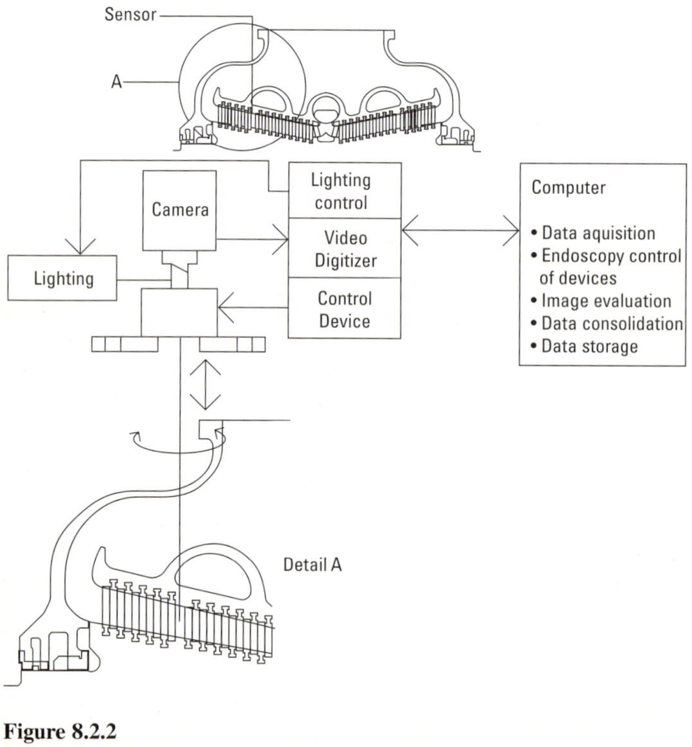

Figure 8.2.2
Borescopical Examination of Blading at closed Turbine During Turning-Gear Operation

without dismantling the turbine. The image is conducted to the camera by fiber optics. A computer will be used to evaluate the photographs in order to detect any changes in the blade surface as compared to the original condition. This technology makes it possible to perform an inspection of the turbine blading in a considerably shorter period of time and reduces the likelihood of human error. For future applications fiber-optics cables only a few millimeters in diameter will provide the possibility to inspect HP turbine blading also. A major contribution to the application of modern methods for machine diagnosis came with the introduction of new sensor techniques.

A further example of an off-line inspection method has been developed for a generator air-gap analysis device to inspect the condition of the generator in the rotor region. This device makes it possible to check the condition of the generator and to perform minor operations in this region without having to remove the rotor. The analysis device consists of an annular video cage which houses the video and drive system for positioning the inspection device in the air gap of the generator and the control equipment for moving the device into the generator. The device makes it possible to perform the following inspections and activities:

- ▷ Inspection of the tooth-tip and key surface,
- ▷ Recovering loose parts/foreign bodies, taking wipe samples, sucking off deposits,
- ▷ Measurement for determining the residual spring travel of the slot-top ripple spring,
- ▷ Sonic testing of the rotor retaining rings,
- ▷ Localizing defective insulation in the laminated core,
- ▷ Borescope inspection of the cooling slits and slot insulation,
- ▷ Localization of partial discharges.

The device makes it possible to gather information on the condition of the generator and to perform minor corrective actions via the generator manhole or after unbolting a bearing cover, without having to remove the rotor. This saves time and money.

An analysis of significant disturbances performed on a steam turbine (Table 8.2.3) shows that a large number of disturbances occur within a time frame in which, provided diagnostics systems are installed, it would be possible to recognize incipient faults. Diagnostics systems (such as DIGEST) are also used to assess the status of components after certain operating periods as a function of the load conditions under which the components have been run. It is then possible to deliberately avoid certain steam temperature ranges, e.g. markedly lower reheat temperatures, in order to prevent the occurrence of droplet impact erosion, or the degree of droplet impact erosion already sustained can be determined by calculating the hazard potential and the number of operating hours run at the corresponding temperatures. There are a number of examples that work along these lines.

Table 8.2.3 Significant Malfunctions of Steam Turbines

Malfunction		Time Spa	Detectability Diganostics Module	Indication (+) Avoidance of further Deterioration (++)
Changes in the blading due to deposits, erosion		w	t, i	+
Impact of foreign matter on HP/IP/LP turbine		h	t, i	+
Ventilation heating of HP/IP/LP turbine		min	va, t, v	++
Return flow from extractions		s	va, v	+
Restricted expansion		h	v, b	++
Leakage in area of steam admission to HP/IP turbine		d	i	+
Shaft distortion (line of shafting)		min	v, b	+
Rubbing of shaft seals		min	v	+ (++)
Damaged shaft seal in HP/IP/LP turbine		min	v, t	+
Deteriorated bearing oil condition		min	i	++
Dynamic bearing load too high		s	b, v	+
Oil fouling		m	b	++
Bearing clearance in turbine/generator bearings		min	b, v	+
Increased axial thrust		s	b	++
Faulty bearing pedestal mounting		h	v	+
Rotating blade break in HP/IP/LP turbine		s	v	+
Journal bearing temperature increase		s	b, v	++
Increased transversal shaft crack		m	v	+
Cooling water leakage	condenser	min	c	+
Air ingress	condenser	min	c	+
Change in heat transfer	condenser	h	c	+
Change in condensate conductivity	condenser	min	c	++
Changes of thermodynamic condition (pressures, temperatures, mass flow)	condenser	min	t	+ (++)
HP/IP casing distortion		min	v	+ (++)
Increased vibration of shaft and bearings		min	v	+ (++)
Increased vibration last stage blade		s	bessi	+ (++)

Legend for time span

s = seconds, min = minutes, h = hours, y = years, d = days, w = week, m = months

Legend for diagnostics modules

va = valves, v = vibration module, t = thermodynamics, ts = service life, bessi = blade vibration, c = condenser, b = bearings and oil system, i = inspection (closed turbine)

The tools available today already make it possible to recognize incipient disturbances by means of DIGEST modules in on-line operation backed up by targeted measurements performed off-line with the turbine generator set shut-down. Experience with faults to date indicates that it is possible to recognize the symptoms of a large number of incipient disturbances by means of dedicated analyses. The range of faults covered is being constantly extended thanks to communication between power plant operators world

Table 8.2.4
Ranges of Various Achievable Savings of Power Plants

Potential savings	Cost savings in Mio DM 400 MW fossil-fueled power plant	450 MW combined cycle power plant	Nuclear Power Station DWR 1300 MW
Cost savings for a major inspection or appropriate module inspections	2 – 3	2 – 3	2 – 5
Cost of repairing major damage	0.5 – 5	0.5 – 3	0,5 – 10
Avoidance of a 4-week outage	8	6	50 – 60
Purchase of electricity from external source for 4 weeks in a plant-specific operation mode	10 – 15	12 – 20	60 – 80
Avoidance of deterioration of efficiency by 0.25 percentage points	5 – 10	4 – 9	20 – 30
Increasing availability by 0.5 percent	5 – 8	3 – 6	30 – 40

wide and the evaluation of faults reported by our customers and in the literature. Fault simulation programs in the DIGEST system make it possible to train operating personnel on computers in the field to enable them to recognize faults and their effects faster and more reliably. Service arrangements are available for continuous upkeep and adaptation. Systematic evaluation of faults reported continuously extends the knowledge base available in the expert systems and thus also makes the diagnoses generated even more reliable. Knowledge of the actual condition of the turbine generator set, as provided by DIGEST on-line in combination with off-line inspection of the machine, form a sound basis for:

- ▷ Recognizing incipient disturbances,
- ▷ Deciding on further action to remedy the disturbance and keep the turbine generator set in operation,
- ▷ Giving instructions for future service inspections via an interface to the operations management system,
- ▷ Avoiding secondary damage and protracted and expensive power plant outages, possibly necessitating the purchase of power from external sources.

Potential cost savings achievable at power plants thanks to the absence of secondary damages and thermodynamic optimization are quantified in Table 8.2.4. Diagnostics systems are useful tools by accomplishing cost savings in power plants.

Table 8.2.5 Damage Avoidance by Means of RF Monitoring

Typ of Faults	Damage Avoided	Outage Time Avoided
Poor contact at the carbon brushes of the shaft grounding	1. Gear unit damage at the main oil pump due to ground fault in the stator windings 2. Bearing damage due to sparking in the bearing shell	7 - 10 days 20 - 30 days
Detachment of a grounding connection at the generator circuit rings	Loss of integrity of circuit ring insulation, i.e. single-phase short circuiting of the stator windings	30 days
Faulty grading layer in a high-voltage bushing	Damage to the bushing, i.e. single-phase short circuit	5 days
Solid conductor break in the voltage transformer winding	Destruction or explosion of the transformer	3 days
Internal high-voltage discharge in one of the generator transformers	Damage to windings in the generator-transformer	5 - 25 days

As experience with the use of DIGEST modules shows, taking the example of the RF module in Table 8.2.5, use of the module for analysis of disturbances made it possible to pre-empt major secondary damage and to avoid a long plant outage.

A further example (Figure 8.2.6) shows what kind of secondary damage and plant outages can occur if the symptoms of deterioration of a turbine generator set bearing are not recognized in time and bearing damage actually occurs. The resultant savings achieved by the avoidance of plant outage and any purchase of electricity from external plants can be calculated by Table 8.0.

Detection of Incipient Faults of Bearings with the Diagnostics Module →

Detection of bearing problems by:
Trending and Analysis

Preventing of secondary damage of the bearing by:
Bearing shell replacement
Plant Outage 2-3 days

Without Diagostics Module
Secondary Damage →

Postulated failure of the bearing

Bearing shell replacement
Machining of bearing journal
Replacement and renewing of labyrinths in glands and shaft
Outage duration 3-4 weeks

Figure 8.2.6 Economic Benefit by a Diagnostics Module Bearing

8.3 Monitoring of Service Life Expenditure

Evaluation of turbine plant availability reveals that the probability of outages increases with plant age. The primary service-related factors are wear, corrosion, erosion, deposits and deformation of components sustained by the individual components during the course of plant operation. The service lifetime of a steam turbine generator [30] can be theoretically divided into the following phases:

- ▷ Initial operating period with predominantly time-independent disturbances or failures (Phase 1),
- ▷ Main operating period with foreseeable and approximately constant wear and tear of components (Phase 2) and
- ▷ Period of increased life expenditure due to increasing material fatigue.

The above influences are depicted in Figure 8.3.0. At the left-hand side of this Figure, time-independent factors are predominant, time-dependent influences prevail at the right-hand side of the Figure. Determining residual service life is therefore a predominant factor in terms of plant availability. Monitoring residual service life can be accomplished with the turbine stress controller or by implementing DIGEST as an integral module. Continual calculation of component residual service life based on plant operating data permits

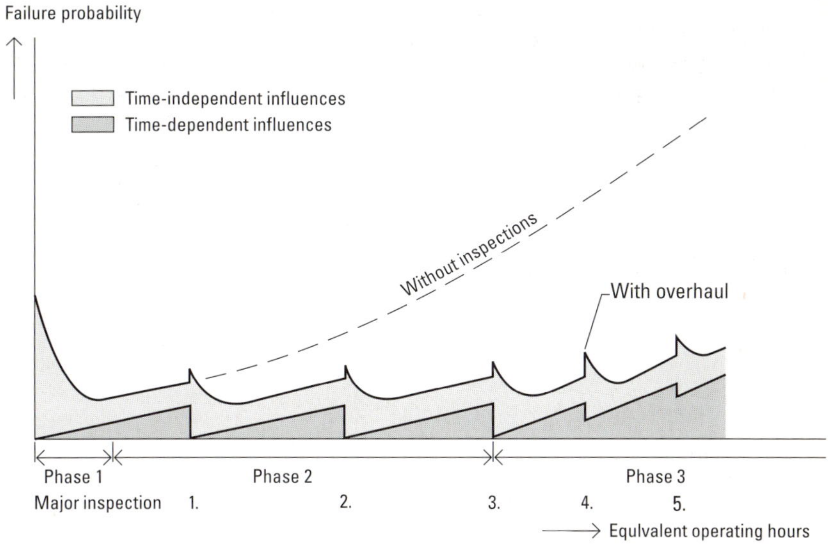

Figure 8.3.0 Failure Probability of a Unit as a Function of its Service Period

optimization of plant operation planning and turbine generator maintenance strategies as follows:

- ▷ Matching of load changes necessitated by the grid or unit optimization in terms of the available duration of turbine generator service,
- ▷ Optimization of start-up processes by selection of various loading limits relevant to residual service life, with service life utilization, to minimize fuel costs,
- ▷ Planning of targeted component tests and inspections on the basis of plant residual service life and the specific service life consumption of individual components.

Monitoring of residual service life permits determination of material fatigue, i.e. accumulated low cycle and creep fatigue and the degree of accumulated stress crack growth of the turbine shaft in question to prevent jeopardizing plant safety and reliability and also permit, performance of inspections geared to ensuring high availability. In light of the vast potential for damage, conceivable savings can be realized when compared to plants that are not equipped to evaluate these criteria, is a factor of great importance; however, this cannot be quantitatively determined in advance.

8.4 Maintenance

The economic operation of steam turbine installations depends essentially on their thermal efficiency and availability. In the past, regular maintenance inspections have proven to be an important element in ensuring safe and economically optimum operation [30]. The all-round assessment of the condition of steam turbine generator sets calls for on-line monitoring and the use of diagnostics systems but also requires the performance of additional measurements on the turbine generator set during shut-down and in certain situation, a fact-finding inspection which may necessitate opening up parts of the unit. Certain findings can only be made during inspections on the open machine. The scope of maintenance inspections is generally geared towards the requirements as set out in the recommendations of the power utilities associations (in Germany the Technische Vereinigung der Grosskraftwerksbetreiber, VGB). Maintenance inspections may be divided by scope into:

- ▷ Minor maintenance inspections,
- ▷ Intermediate maintenance inspections,
- ▷ Major maintenance inspections,
- ▷ Modular maintenance inspections.

Modular inspections are predominately used for steam turbine generator sets in nuclear power plants. In the above-mentioned inspections the scope of measurements, examinations and recording to be performed ranges from straight checks and examinations (minor maintenance inspections) to opening up of the turbine and generator casings and dismantling of the major components (major maintenance inspections), while modular maintenance inspections involve the detailed inspection of the successive turbine casings (HP, IP, LP) at staggered intervals according to various assessment criteria. A maintenance inspection in which all the turbine casings of a large steam turbine generator set are opened up has a major impact on the power plants operating costs and makes a significant inroad into plant availability. At today's state-of-the-art and with modern operating modes, maintenance inspections are no longer so important as an opportunity to restore the turbine to higher efficiencies, unless measurements and analyses already performed in service give reason to suspect the presence of wear or damage.

For this reason, the plant operators-call for optimum operation also includes the desire to further extend the intervals between maintenance inspections and to replace the traditional strategy of time-schedule-based planning of maintenance inspections by other approaches that allow the plant to stay on-line for longer periods. Figure 8.4.0 shows the strategy proposed by the VGB with a view to achieving these aims. In order to determine the maintenance inspection interval, it is recommended that one proceeds on the principle of equivalent hours of operation, which also takes the effects of the number of starts into account. The equivalent hours of operation are defined by the following equation: $T_e = T_H + n_s \cdot T_s$

T_e = Equivalent hours of operation,

T_H = Actual number of hours in operation,

n_s = Number of starts, not differentiated according to warm or cold starts,

T_s = Number of hours of operation to be calculated for a start (20 to 30 hours).

Generator systems are also subject to maintenance inspections at certain intervals. Power station operators try to establish for turbines and generators

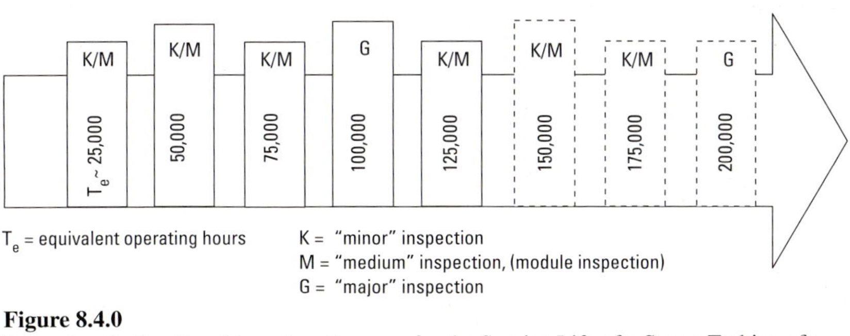

Figure 8.4.0
Example of a Baseline Inspection Program for the Service Life of a Steam Turbine of a Fossil-Fired Power Plant

similar inspection cycles. After a relatively short time of approx. 15,000 equivalent operating hours the initial inspection is scheduled. This involves a thorough crawl through inspection performed with modern tools (video technique) by specialists, examining also the entire air-gap area. Under these circumstances for new generators up to 400 MVA the removal of the rotors is not required. This might also be practiced on larger generators in the future. Following such an initial inspection, checks and cleaning activities are performed as minor maintenance inspections during the course of, e.g. boiler inspection cycles after about 25,000 equivalent operating hours (as an approximate value) of operation. The first major maintenance inspection (with removal of the rotor) is then to be performed following the initial inspection after an additional T_e of about 50,000 hrs. At this time all wearing parts and the sealing elements subject to aging are replaced as a preventive measure. These intervals between maintenance inspections – the minor maintenance inspection after about 25,000 hours and the major maintenance inspection after every 50,000 equivalent hours of operation – are continued until the findings make it necessary to change them. In summary, the modern maintenance philosophy is based on the principle:

No major maintenance when measured values and observations (analyses) are regular and without changes and when there are no other negative findings. If the measured values approach the limiting values or a diagnostics module gives an indication of on-service failures, the date and scope of a maintenance must be determined taking into consideration the evaluation of the various parameters.

Increasingly use of diagnostics systems will help to extend the intervals between major maintenance inspections of turbine generator sets and eventually to introduce the concept of the status-oriented maintenance. The objective of all preventive maintenance strategies is to operate capital-intensive plants up to the end of their intended service life with a minimum of

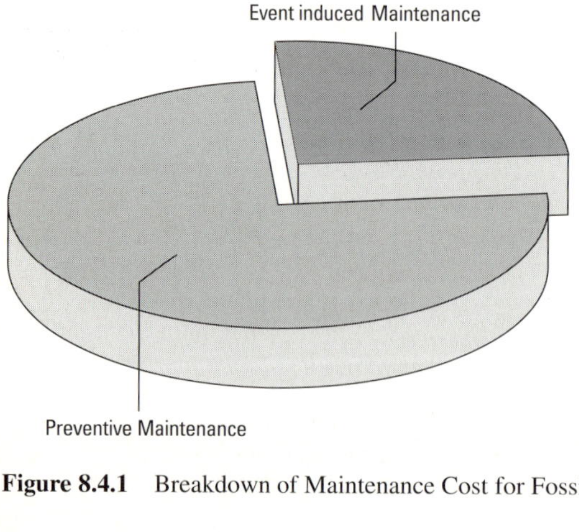

Figure 8.4.1 Breakdown of Maintenance Cost for Fossil-Fired Power Plants [31]

cost, optimum cost-effectiveness and maximum availability. Because of the degree of sophistication achieved in terms of mechanical and functional design and in view of quality-controlled manufacturing and assembly methods, modern turbine generator sets in toady's power plants and the application of diagnostics systems are particularly suited to the application of the status-oriented maintenance strategy. Whereas the common practice used to date involved a periodic strategy based on equivalent operating hours, the future maintenance strategy concentrates on plant condition monitoring diagnostics and inspections preferably on the closed machine under standstill condition (off-line).

Analysis of availability statistics show that shut-downs of turbine generator sets caused by sudden faults are restricted to a relatively small availability percentage. The relatively low forced outage rate is commonly achieved through application of the maintenance strategies previously described. The cost of the preventive maintenance strategy in thermal power stations is, however, much higher than the cost incurred due to sudden faults and damage of major components, even if the plant being suddenly shut-down with long downtimes. Figure 8.4.1, [31] gives an overview of the maintenance costs with regard to the costs which are event induced. A further reduction in the proportion of costs caused by faults and damage by increasing the amount of maintenance is hardly possible, nor is it economically acceptable.

A cost reduction can only be achieved by the status oriented maintenance strategy using diagnostics systems, providing efficient analyses and diagnoses with reliable information and user-friendly processing of the results. In view of achieving a high cost-saving potential by further improving plant availability and reducing maintenance costs, the development of DIGEST has been given top priority and constantly adapted to advances in the state-of-the-art. Avoiding a major maintenance makes it possible to achieve considerably high savings: The savings achievable by an optimized maintenance strategy with the resultant increase of availability can be calculated with the formulas given in the Table 8.0. Results for various savings in this field for different types of power plants are shown in Table 8.2.4.

Analysis of measured data and the application of smart sensors on operating units with our on-line diagnostics system DIGEST allows us to make far-reaching judgments on the unit's condition and to anticipate component failures. On-line diagnosis is backed up by inspections of the machines in the shut-down condition on the basis of sophisticated techniques. Under this concept, disruptions, damage or uneconomical operation can be detected at a very early stage and the cost-effectiveness of the plant can be increased as a result.

Implementation of a condition-based maintenance strategy is only possible if knowledge of the turbine generator equipment condition is available. The individual phases of reducing the overall maintenance effort brought about by inspections and implementation of diagnostic systems are depicted in Figure 8.2.4 in terms of major maintenance inspection scope to give a simplified, qualitative illustration of the concept's benefits. Inspection findings on components constitute the primary source of information for defining the type and scope of maintenance and repair measures actually needed.

If it is possible to procure information on the condition of the turbine plant by means other than the fact-finding inspection performed during a major inspection, the transition can be made to a condition-based maintenance and

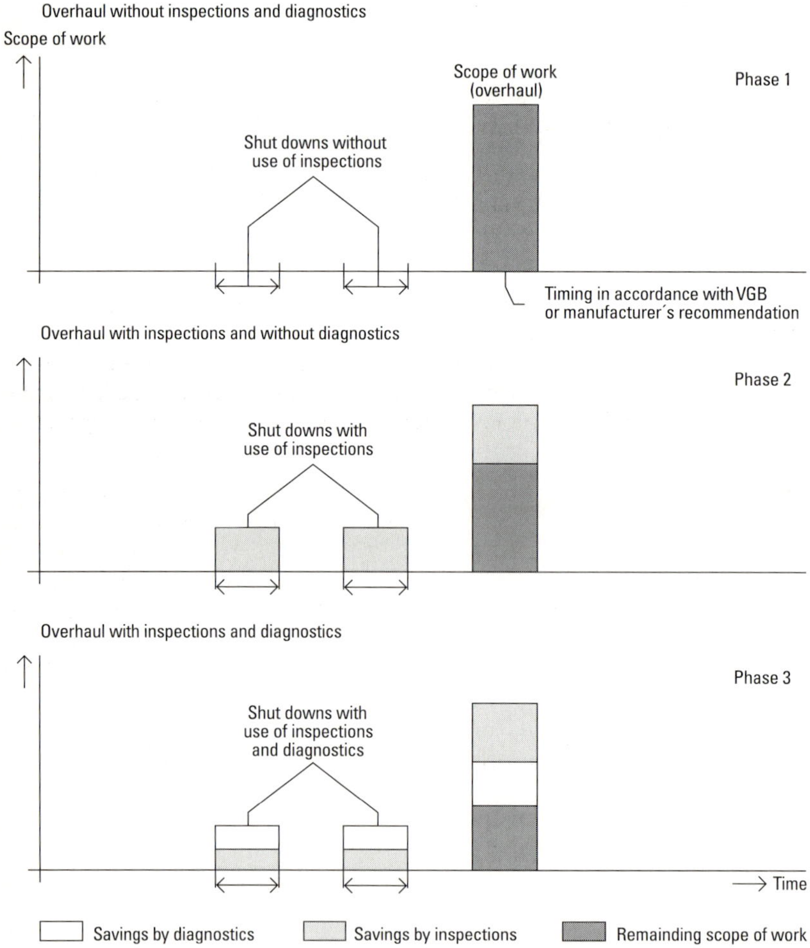

Figure 8.4.2
Maintenance Strategy and Savings by Application of Inspections and Diagnostics

inspection planning strategy. Here the objective is to schedule the requisite scope of work during an overhaul on the basis of equipment condition in advance. Figure 8.4.2 illustrates a procedure intended to effectively implement inspection and diagnostics activities as well as condition-based maintenance. This Figure shows savings on a qualitative basis achieved by reducing the volume of an extended major maintenance as follows:

Phase 1

Major maintenance inspections are performed on the basis of equivalent operating hours. Outages other than major maintenance inspections are not utilized for performing scheduled inspections.

Phase 2

Targeted implementation of modern inspection methods reduces the major maintenance inspection effort. Major maintenance inspection measures dealing with individual items and subassemblies whose condition can be evaluated by inspection are only performed as required.

Phase 3

Implementation of diagnostics systems permits determination of the current condition of the turbine plant to a certain extent. Results of the diagnosis permit establishing of maintenance and inspection efforts on the basis of plant equipment condition which, in turn, reduces the risk of sudden plant shut-downs. Reduction of scope of requisite inspections and major inspection efforts is the result in the event that there are no negative findings.

The objective of phase 3 can already be achieved to a large extent by implementing on and off-line diagnosis and performing limited scope inspections. The DIGEST diagnostics system, developed for steam turbine plants, permits analyses and diagnoses with regard to proper functioning of the steam turbine generator set, optimum operation of the plant and provides a solid basis for introducing status-oriented inspection strategies by means of which the high costs of preventive maintenance can be reduced. These aspects of the diagnostics system DIGEST result in an improvement of the economy of the power plants.

9 Service and Experience with DIGEST

9.1 Service, Basic Concept

Evaluation of data and analyses produced by the diagnostics system is of great use to maintenance and service. The manufacturers of turbine generators gear their diagnostics system to the respective details of design and plant operation. Qualified personnel are familiar with operation of their own systems, thus optimum benefit is achieved for the customer. With this foundation, it is possible to modem data to the manufacturer for fault evaluation and consultancy by their engineering staff.

Siemens/KWU has developed an operations management system to provide computer support for the economic operation of power plants. This system covers all fields of activity within the power plant and also caters to the aspects of optimum servicing and maintenance. The requisite interfaces to DIGEST can be realized.

In view of enhancing plant availability and achieving the highest possible thermal efficiency, data from the plant on a scale to be agreed on with the plant operator, should be submitted to the manufacturer, too, in order to enable him to propose improvements for enhancing efficiency on the basis of comparisons with other similar plants. Siemens/KWU has also developed strategies for this aspect of plant monitoring and covers as an manufacturer of steam turbine generator sets and I&C equipment a broad scope of services in this field.

9.2 Service, System Upkeep

Experience gained from the use of diagnostics systems shows that, in the interest of achieving optimum benefit, customer-side system upkeep as appropriate to the scope and type of the diagnostics modules is necessary. System upkeep should include knowledge of the process-engineering background and the ability to look after the hardware. Furthermore, close cooperation between the system coordinators on the customer's and the manufacturer's side is necessary. The benefits expected of diagnostics systems can only be fully exploited in the long-term by close cooperation between both parties. Servicing by the manufacturer is advisable for diagnostics systems already installed. Siemens/KWU covers the scope of diagnostics upkeep required in the various fields:

- ▷ Instrumentation & Control systems (sensors, bus systems, I&C equipment, etc.),
- ▷ Hardware (computers),
- ▷ Software (user programs, basic software, operations management system),
- ▷ Technology (expanding the knowledge base for more comprehensive diagnoses and interpretation of faults via modem or at site),
- ▷ Proposition of product oriented maintenance concepts,
- ▷ Continued development and update of installed diagnostics systems to the state-of-the-art,
- ▷ Possibility of adding on new developed modules step-by-step,
- ▷ Personnel training for fault analyses.

An approach should be defined for updating of and adding to the knowledge bases and data banks by the manufacturer and for incorporating suggestions by the power plant operator. This will involve solving problems in the various technical fields, deciding on how to interpret incipient disturbances in components and systems, extending the knowledge bases, and long-term upkeep, for instance adaptation of the hardware and software to progress made in these fields. Siemens/KWU has developed a strategy for the various services that can be tailored to the individual customer's requirements. It provides for installation of a modem for communication with the manufacturer of the respective diagnostics module. The data from the separate modules of the diagnostics system can be fed into the customer's operations management system or can be stored in the DIGEST memory so that important maintenance information is available at all times.

9.3 Experience with DIGEST Modules

The processing of parameter readings and signals from the plant by means of on-line diagnostics systems facilitates comprehensive analyses of the plant operating status and incipient changes with a view to recognizing faults, damage and uneconomic plant operation as early as possible. Examples of cost savings which can be achieved through the use of DIGEST, are given in chapter 8.

The modularization of DIGEST provides for high versatility for use in all types of power plant, including backfitting in plants already in operation. DIGEST lends itself to extension step by step. Once a basis module with the associated software and hardware is in use in any application, further modules can be added as required.

Table 9.3.0 DIGEST Modules on Order and in Operation (Date of Compilation June 1996)

Module	Chapter	Number on order per unit	First Module in Operation
Vibration	7.1	14	1994
Service Life Evaluation	7.2	9	1988
Thermodynamics – Unit Commitment	7.3.1/2 7.3.3	25 1	1990 – *)
Windage	7.4	–	– **)
Blade Vibration	7.5	3	1991
Condenser	7.6	3	1994
Bar Temperature	7.7	25	1984
RF Monitor – Diagnosis (Stage 1) – Diagnosis (Stage 2)	7.8.1 7.8.2 7.8.2	24 2 1	1985 1995 1998
Hydrogen Sealing	7.9	1	1993
High Humidity	7.10	1	1996
Gas Turbine	7.12	66	1994
Feed Water Chemistry	7.13.5	1	1997

*) order under negotiation, **) under test

Experience gained from use of several DIGEST modules have already proved this system to be economically. Table 9.3.0 gives a review of DIGEST modules on order and in service:

Most of these modules have a sufficiently long and successful service history of bringing faults and the results of optimizations to the attention of the operators and thus preventing serious damage to the plant and long costly outages and a more economical operation.

Broad-based diagnoses of the steam turbine generator set for the purpose of status-oriented overhauls to a considerable high degree are achievable with the DIGEST modules and the associated tools for off-line diagnostics even today.

Analyses of thermodynamic deviations from design parameters, avoidance of secondary damages and optimization of maintenance strategies reveal that considerable savings can be achieved by the application of a comprehensively equipped DIGEST, particularly with regard to the long service life of today's power plants. As experience shows, the saving potential achievable by DIGEST far exceeds the capital investment of this system.

Literature

[1] Zörner, W.: Steam Turbines for Power Plants Employing Advanced Steam Conditions. Contribution of the 10th CEPSI Christchurch, New Zealand 1994: Siemens AG, 1995

[2] Markus v. H.: Turbogenerators. Siemens Power Generation: Siemens AG, 1993

[3] König, N.; Sill, U.; Wetzl, R.; Deckart, K.: Instrumentation & Control for Power Generation. Power Gen. Asia '93 Sept. 13–15 Singapore, Volume II pp. 491–514

[4] Diegel, D.: Digitale Leittechnik am Turbosatz. Siemens Power Generation: Siemens AG, 1990

[5] Ziebart, H.; Gobrecht, E.: Rechnergestützte Lebensdauerüberwachung von Dampfturbosätzen. VDI-Bericht Nr. 868, 1991

[6] Kürten, H.; Termühlen, H.: Design and Operating Criteria of Fossil-Fueled Power Plants in Europe. ASME Paper No. 76-JPGC-Pwr-1

[7] Lellek, H.: TRS System 82 for Steam Turbine Generator Instrumentation and Control. Siemens Power Generation: Siemens AG, 1988

[8] Zörner, W.; Müller, H.; Andreae, K.-H.; Emshoff, H.: Diagnostics System for Monitoring the Steam Turbine Generator Set. VGB Kraftwerkstechnik 06/1991 pp. 487–498: VGB Kraftwerkstechnik Essen

[9] Emshoff, H.; Müller, H.; Rehbold, R.: Field Experience with a Fuzzy Expert System for Turbine Generator Diagnostics. Third International Conference on Industrial Fuzzy Control & Intelligent System Houston Texas: Addision-Wesly, 93

[10] Müller, H.; Scheil, E.; Einsatz eines Fuzzy-Basierenden Expertensystems zur Diagnose an Turbogeneratoren. Siemens Power Generation: Siemens AG, 1988

[11] Lippmann R. P.: An Introduction to Computing with Neural Nets, IEEE-ASSP Magazine pp. 4–22, April 1987

[12] Holland, J.: The Creation of the Genetic Algorithm. Handbook of G. A.: Van Nostrand Reinhold, New York 1991

[13] Erler, H.; Lausterer, R.; Zörner, W.: Control Room Instrumentation, An Expert Approach to Power Plant Optimization. Power, Special Report, Modern Power Systems, October 1992

[14] Andreae, K.H.; et al.: Rechnergestützte Schwingungsüberwachung und Diagnose an Kraftwerksturbosätzen VDI-Bericht Nr. 846 1990

[15] Ziebarth, H.; Baumgartner, R. J.: Early Detection of Cross-Sectional Rotor Cracks by Turbine Shaft Vibration Monitoring Techniques, Contributed by the Society of Mechanical Engineers ASME/IEE P.G.C. Conference October 4–8, 1981 St. Louis Miss.

[16] Balling, L.; Joyce, J.S.; Rukes, B.: Thee New Generation of Advanced GUD Combined-Cycle Power Plants, supplemented version of the Publication presented to Power-Gen Europe, Amsterdam, May 16, 1995. Siemens Power Generation: Siemens AG, 1995

[17] Zörner, W.: The Diagnostics System "DIGEST" Optimizes the Operation of Combined-Cycle Power Plants. Presented at I.P.G.C. Atlanta GA, October 18 – 22, 1992

[18] Hillermeier, C.; et al.; Kombination von Neuronalen Netzen und Genetischen Algorithmen zur Kraftwerkseinsatzplanung. Tagungsband zum GI-Workshop "Fuzzy-Neuro-System 95" Darmstadt, Seite 257 – 264

[19] Uicker J.; et al.: Windage Effect on HP Sections of Reheat Steam Turbines. Presented at the 1994 International Joint Power Generation Conference IJPG, Published by the ASME P.D.

[20] Bergmann, D.; et al.: High Temperature Control in High Backpressure LP-Turbines. American Power Conference Chicago, Illinois Apr. 22 – 24 1985

[21] Gloger, M.; et al.: Blade Vibration Information System "BeSSI" for Power Plant Operation. Presented at the IJPGC 95 in Minneapolis

[22] Zörner, W.; Drosdziok, A.: Functional and Thermodynamic Control of Steam Turbine Condensers and their peripheral Systems. Presented at ASME-IEEE Power Generation Conference Philadelphia Penns. Sept. 25 – 29, 1988

[23] Hachfeld, D.; Sauer, H.: Erfahrungen zum Einsatz der "DIGEST" Kondensator-analyse im KKW Grohnde. VDI Bericht Nr. 1160, 1994

[24] Emshoff, H.; Krug, M.: Water-Cooled Stator Windings of Turbogenerators. Computerized Monitoring of Water Temperature-Rise for Early Detection of Over-heating due to Flow Restrictions in the Stator Winding. Presented at the International Conference on Large High Voltage Systems, Paris 1986, 27th August – 4th September

[25] Grünewald, P.; Weidner, J.: Possibilities and Experience with Off- and On-Line Diagnosis Turbine Generator Stator Winding-Insulations. Publication CIGRE 28. Aug. – 03. Sept. 1994, Paris Session

[26] Weidner, J.; Grünewald, P.; Schlager, U.: Turbine Generator On-Line Monitoring and Diagnostics System. EPRI-Presentation, Maintenance Workshop Scottsdale, AZ Dez. 10 –12, 1991

[27] Bischoff-Beiermann, B.; Bode, A.; Werner, K.: On-Line Gas Turbine Diagnosis. Siemens Power Journal 3/94: Siemens AG, 1988

[28] Köncke, K.; Zörner, W.: Recently Developed On-Line Diagnostics Systems for Steam and Gas Turbines. Power-Gen Asia Singapore 1993 Sept. 13 –15, Volume II pp. 43 – 56

[29] Zörner, W.; Denkes R.: The "DIGEST" On-Line Diagnostics System Reduces Costs in Power Stations. Joint Power Generation Conference San Diego 07.10.–10.10.91

[30] Münstermann, H.; Morgan J.: Service Concept for Turbine Generators in West Germany. Presented at the Joint Power Generation Conference, Philadelphia, PA. Sept. 1988

[31] Jäger, G.; Dibelius, G.: Automatisierte Zustandsdiagnose zur Optimalen Instand-haltung von Dampfturbosätzen: VGB-Kraftwerkstechnik, Vol. 68, Dez.

University of Massachusetts Lowell Libraries